THE FIRE BLITZ

Burning Down Japan

THE FIRE BLITZ

Burning Down Japan

John D. Beatty

•JDB•COMMUNICATIONS,•LLC•

JDB COMMUNICATIONS, LLC
WEST ALLIS, WISCONSIN

First Paperback Edition ISBN: 979-8-9860169-6-2
First E-Book Edition ISBN: 979-8-9860169-7-9

For Paul Heim and Dale Roethig:
if only you guys could see this now…

Contents

Introduction

In December 1941, Japan attacked Pearl Harbor, and the US entered World War Two.

A bunch of island battles followed.

In August 1945, the US dropped two atomic bombs on Japan, ending World War Two.

***

That has been the narrative for the Pacific War and the end of World War Two since 1945. There was a great deal more, more even than the island battles from 1941 to 1945, but the atomic bombs dropped on Hiroshima and Nagasaki have eclipsed everything that happened in between Pearl Harbor and Nagasaki, including why and how the conflict *actually* ended.

The atomic bombs were only *part* of the ending.

The only aircraft that could carry those atom bombs was the Boeing B-29, called the Superfortress. When the B-29 first flew in September 1942, neither its most generous observers nor its builders thought of it as a practical platform. Yet, a desperate United States Army Air Forces (USAAF) general ordered over two hundred more planes based on buggy and hard-to-fly prototypes, and didn't change his mind even after the second aircraft crashed, killing Boeing's best test pilot.

The USAAF intended for the Superfortress, which was *becoming* the most expensive weapons system of the war, to show that air power alone could defeat an enemy—*any* enemy—without costly land attacks. By early March 1945, after nine months of bombing missions from India, China and the Marianas, the USAAF still could not show that the big, spendy plane could bomb Japan *effectively* at all. Less than a year after it became operational, the technically advanced yet cantankerous airplane was at a crossroads. It could either lead the world in air power, or play a secondary, supporting role to land and naval forces and other, cheaper airplanes.

The atomic bomb, remember, didn't *exist* until July 1945.

The first demonstration of *how* to use the B-29s *effectively* would take place in a perilous and *horribly* destructive air campaign starting in March 1945 that some called *The Fire Blitz*. How that campaign came about and the events that *followed* it is the heart of this story.

Preludes

18 April 1942

Some of us did feel a little strange when we noticed that one of the planes dropped something that caused some smoke when it reached the ground...[1]

Aikawa Takaaki

At 7:28 that morning, *Nitto Maru* (picket boat Number 23 of the Imperial Japanese Navy's (IJN) 22nd Picket Squadron) transmitted a message stating that it had sighted a force of three American aircraft carriers 720 miles east of the Japanese Home Islands. Tokyo *probably* tried to get more information[*] but *Nitto Maru* went silent after that message went out.[2]

If the report was at all valid, given the position of *Nitto Maru* and the assumed range of carrier-based aircraft, the authorities in Japan believed an attack would *not* happen until the next day, if at *all*. Still, erring on the side of caution (something *had* to have caused *Nitto Maru*'s sudden silence), the IJN alerted all their available aircraft in the Home Islands—90 fighters and 116 bombers—to the *possibility* of an attack in *no less* than 24 hours. The IJN also ordered six heavy cruisers and ten destroyers out of the Yokosuka Naval Base to search for and attack the

---

[*] Sources are unclear.

American force, if there *was* one...but there had to have been *something*. Japanese authorities *knew* an American attack would *not* come until their aircraft carriers were within three hundred miles of the Japanese coast, if it came at all. It is not clear if the IJN alerted their Imperial Japanese Army (IJA) counterparts.[3]

At about 9:00 that Saturday morning, air defense authorities in Tokyo held a scheduled air raid drill, beginning with a whooping siren and the release of barrage balloons to their optimal height. Since it was not a workday for many, Tokyoites enjoyed the spectacle of the fire brigades with their shiny equipment fighting fires they set in stacks of brushwood. The spectators ignored the officious fire wardens, who told them to just seek shelter as they would in a *real* air raid.

At 9:45, a patrol plane reported an American two-engine bomber east of Japan, flying west. Japanese authorities did not believe this report—there were no enemy bases nearby and aircraft carriers did not *carry* two-engine bombers. This was a *known fact*. Therefore, this report *must* have been a mistaken ID of a Japanese aircraft.

At about 11:30, Tojo Hideki, Prime Minister *and* War Minister of Japan, tried to land his plane at the Mito Aviation School in the Ibaraki Prefecture, in the north of Kanto Island. He found it odd that several unidentified twin-engine bombers were in the same airspace as his. Earlier that morning, he received information about a *report* of an American naval force being far away. He felt certain that it would be safe to perform an inspection of the IJA flying school, even *if* the report were true. His secretary *thought* he recognized an *American* in one cockpit. Tojo's reaction to this is unknown, but was probably dismissive.

At about noon, the authorities sounded the all clear in Tokyo; the IJA reeled in the barrage balloons.[4]

At 12:30, James H. "Jimmy" Dolittle and twelve of his Raiders were over Tokyo in their B-25 bombers, dropping bombs. Tokyo radio station JOAK went off the air immediately; crowds, seeing the planes, waved at them enthusiastically. At 12:34, the Tokyo air raid sirens sounded once again, and some feeble antiaircraft fire began but soon ceased. One source credits a single interceptor sent aloft. The Japanese shot down no Raiders.[5]

Smoke rising from targets hit during the Dolittle Raid, April 18, 1942 (US Air Force).

The Dolittle Raid killed no more than fifty Japanese and injured an unknown number more, and damaged or destroyed an unclear number of buildings. But the Dolittle Raid shocked the IJA, as they were responsible for defending the Home Islands. The IJN, which only had responsibility for its naval facilities and ships, nonetheless sought to augment its defenses. The very idea that the sacred homeland had been attacked had been unthinkable just the day before. Now, great waves of recrimination washed up in the councils of the samurai. *Someone* was to blame…

9 March 1945

> *In a single night, the history and fate of Tokyo were altered forever.*
> Saotome Katsumoto[6]

The mournful, by-now-familiar yet no less urgent wailing of air-raid sirens roused Tokyo from its all-too-brief Friday night slumber. With fresh news of approaching *B-29-Sans**, the city's air raid wardens—though mildly confused at this *highly* irregular night attack—dutifully rushed into the streets and alleyways to alert the citizens, telling the tardy to seek shelter, warning all to put on their fire hoods. Though the wardens knew this was *not* a drill, they felt certain this would be yet another small-scale raid on the Imperial City of the Meiji.[7]

---

* *San* is a gender-neutral honorific meaning Mister, Miss or Missus. In this case, it translates loosely into *Mr. B*, crediting the B-29's size and payload.

Many residents, working through the night making ammunition and airplane parts, heard the sirens but continued their labors undaunted. Many others, not involved in war-related industries, scurried for the shelters as they always had, but without the urgency of normal, daylight air raids. This *had* to be a *drill*, they assured themselves. Even so, the last air attack on Tokyo, a *very* local but *very* destructive affair, *did* burn out the Ginza with firebombs, even if that shopping district no longer had anything to *buy*...[8]

Mere seconds after the sirens started, searchlights lanced the dark night with their probing, piercing fingers of light, and the antiaircraft guns began firing...

Tokyo, 10 March, 1945 (Wikimedia)

And The Fire Blitz began...

The Leading Characters, Organizations, and Ideas

...An aerial force is a threat to all points within its radius of action, its units operating from their separate bases and converging in mass for the attack on the designated target faster than with any other means so far known. For this reason, air power is a weapon superlatively adapted to offensive operations, because it strikes suddenly and gives the enemy no time to parry the blow by calling up reinforcements.

Giulio Douhet[9]

To understand what led to the birth of strategic aerial bombardment (especially the American version, precision daylight bombardment), the B-29 and the Fire Blitz, it is important to know the players—particularly the Americans—the conflict, concepts and organizations that were midwives to the holocaust.

The Great World War and Aftermath

That 1914-1918 conflict, WWI—or HG Wells' coined phrase, the "war to end all wars"—affected not just Europe, where most of the casualties happened, but the entire world…and changed it in ways no one could have predicted. The body count—ten million and more *without* the influenza deaths—shocked even the most cynical and callous observers. The scale of *industrialized* slaughter—the first major conflict fought on more than one continent with airplanes, radios and steam-powered factories—benumbed the sensibilities of soldier, sailor and non-combatant alike. Reports of the British casualties on the first day of the Somme offensive in July 1916 beggared the imagination.

As if the military casualties weren't enough—with soldiers dying in the mud and filth of the Western Front, the plains of Poland and Russia, the mountains of the Balkans and the Isonzo River lines, the jungles of Africa and the deserts of Sinai—German submarines were torpedoing cargo ships and ocean liners on the high seas without warning, leaving sailors, seamen and civilian passengers to freeze and drown in the Atlantic, the North Sea and the Mediterranean. Their justification was that even a lightly armed cargo ship could sink a thinly armed submarine—true, *but*…. Even civilians in their beds at home weren't safe from marauding warships with long-range guns and high explosive shells, nor from attackers from the sky. The casualties caused by the small bombers of the day—or even the huge airships—were microscopic compared to the damage they caused to morale in Britain and France.

War became truly total. *Civilians* in range of enemy guns, torpedoes, and planes were now merely unfortunate *non-combatants* on a large scale.

At the end of the conflict, politicians signed treaties of peace, understanding, and arms control that, though breathtaking in scope, still relied on the word and intent of all the former belligerents who were, after all, just *people* who need food, shelter and meaningful employment. But the economic shocks of the 1920s and '30s upset the well-meaning intent of the leaders and philosophers who signed the treaties that ultimately were just paper. Treaties could not overcome age-old suspicions and hatreds, nor could they bury them from generation to generation. The factories of even the crudest economies might *not* have been able to make refrigerators for everyone, but they *could* make guns.

Germany, devastated, was left with a democracy without democrats, rudderless and in economic and social chaos, its monarchy abolished. The Weimar Republic that replaced the German Empire tottered along until 1933, when Germany found a strongman preaching an ideology that would lead them back to glory, prosperity, and national pride once again.

All they would have to do was obey their one remaining political party, and believe everything their Leader said.

Japan, a newly declared Great Power, poor and agrarian but industrialized by dragging itself kicking and screaming out of a late medieval society and into the 19th Century, learned a lesson from the WWI blockades of Germany and of Britain: resource autarky was the way of the future. Its hybrid political system (Japan's constitution did not *elect* its chief executive, and gave the military ultimate veto power over policy) tottered on the edge of military dictatorship as *bushido*-infused colonels dreamed of a Japan without politics. The newest guns that they could neither afford to buy or build would empower such a government, assassinating any of those who opposed them along the way. The generals and admirals had the answer: more territory for more resources. And the Japanese people... would, of course, support their one remaining political party and obey what the samurai told them to do. They always *had*. They would *now*.

When the sheepdog no longer has teeth, the wolves find easy pickings. The wolves of Germany and Japan waited for the sheepdogs of Europe, Asia and, most of all America, to shed their teeth and leave their charges to fend for themselves.

Before and during WWI, but growing louder and more influential in the 1920s and '30s, a few military aviators believed airpower could eliminate the need for surface forces altogether. American and British flyers dreamed of replacing the mud of the trenches with the pure air of the heavens, of direct attacks on the *morale* not just of leaders, but of their populations as well. This line of thinking put them constantly at loggerheads with mindsets—military *and* political—that historically drove countries to open-ground based warfare.

Air war promised few casualties, if any...

Giulio Douhet (1869-1930)

Giulio Douhet attended the Military Academy of Modena and the Italian Army commissioned him in the artillery in 1882. After studying at the Polytechnical Institute at Turin, he lectured on military mechanization. When flying machines arrived in Italy, he quickly realized their military potential and became an air power advocate. In the Italo-Turkish War (1911-12) over control of Libya, Italy pioneered the use of aircraft for reconnaissance, artillery spotting, transportation and even bombing. Douhet took command of Italy's aviation assets in Libya in 1912, writing one of the first-ever doctrinal manuals for the deployment of air power. Ever the zealot, Douhet ordered the construction of Caproni bombers without authorization, which got him transferred to the infantry as punishment.

When Italy entered WWI, Douhet wrote articles and letters and made speeches about how aircraft could make up for Italy's poor showing against Austria-Hungary. He proposed a large force of bombers that could break the stalemate in the Alps. Ignored at first, he was finally court-martialed and jailed for a year for the crime of criticizing the government's conduct of the war. From his cell, he continued to advocate air power. Released in 1917, the army appointed Douhet to head the Central Aeronautic Bureau in 1918. After the war and his promotion to general, Douhet published *Command of the Air* (1921) that claimed that air power was the key to future war.

William L. Mitchell (1879-1936)

William Lendrum Mitchell attended Columbian University (formerly Columbian College, later George Washington University) in Washington, DC, but dropped out to join the Army during the Spanish-American War. As the scion of a wealthy and influential Wisconsin family (his father had been a senator; a grandfather was a railroad tycoon), Mitchell was soon on the staff of Arthur MacArthur in the Philippines. His father's influence gained young Mitchell a commission, and he joined the Signal Corps, with which he laid telegraph cables in Alaska. In 1908, Mitchell saw the Wright brothers' demonstration flight at Fort Myer, and soon he was taking flying lessons from Glenn Curtiss. At age 32, the Army selected Mitchell to serve on the General Staff—then its youngest member. When the Aviation Section of the Signal Corps formed in May 1916, Mitchell was appointed its acting head. That July, he became the Chief of the Air Service of the US First Army. April 1917 found Mitchell *en route* to France as an observer, but he soon set up an Aviation Section in Paris, collaborating with the British and French air services. By October 1918, Mitchell was an acting Brigadier General, commanding the American Air Service in France.

The Army disappointed Mitchell by *not* making him head of the Army Air Service. Instead, they assigned him to secondary roles that were soon eliminated. Well respected in aviation circles, he did not share the belief held by many that the Great War *was* the "war to end all wars." Contradicting his superiors, he advocated a separate air force, and loudly criticized those who disagreed with him. As the Navy had largely disbanded their air arm, Mitchell picked a fight with them, too. Given a chance to "prove" that an airplane could sink a ship,[*] Mitchell proved himself correct on 21 July 1921, when he and his bombers sank a demilitarized German battleship, *Ostfriesland*. Mitchell was a prodigious writer, and some of his writings got him in a lot of trouble...*and* court-martialed.

But Mitchell was of private means, and continued speaking and publishing about his ideas, criticizing "The Establishment" at every opportunity. Many listened, others did not.

The Air Corps Tactical School (ACTS)

Also known as the Tactical School, ACTS was a military professional development school for flyers of the US Army, the first such school in the world. Created in 1920 at Langley Field, Virginia, it moved to Maxwell Field, Alabama, in July 1931. In time, the school became the doctrinal development center of the Army Air Corps (AAC), and a preparatory school for flyers aspiring to attend the Command and General Staff College. The motto of the school was *Proficimus More Irretenti*—We Make Progress Unhindered by Custom.

At first, the school taught that pursuit aviation was the most important air operation, comparing its importance to that of infantry to the Army. Gradually, the school began asserting that airpower could strike at vital points deep inside enemy territory, rather than merely targeting an enemy's military forces in a war of attrition. By 1931, the school had become convinced that it was most difficult, if not impossible, to stop a determined air attack.

Devotees of Douhet and Mitchell dominated the faculty, developing a theory of warfare that invoked the superiority of the long-range bomber over all other types of aircraft. Going beyond Mitchell's ideas, they emphasized a doctrine that heavily armed bombers could fight their way to industrial targets in daylight, unescorted by fighters, and with precision bombing, defeating an enemy by destroying its key war production targets, and thus win wars without costly and prolonged ground campaigns. While the idea, originally known as the industrial web theory, was based on tenets of strategic airpower developed by

---

[*] Most naval authorities already believed it possible, but the Navy wanted a more controlled experiment that Mitchell and his men preempted.

Mitchell and Douhet, it rejected Douhet's concept of bombing civil populations to destroy their morale and coerce the enemy's overall *will*.

This doctrine brought them into conflict with the US Army General Staff, which saw airpower as an auxiliary to the ground forces. Despite the poor performance of what few bombers the AAC possessed, the air theorists persisted in their beliefs.

Those who proposed this doctrine viewed war in the abstract. Many of their assertions relied on the use of technology that did not then exist. As such, they had to confess—even apologize for—their inability to offer any conclusive proof of their theories. Yet, they firmly believed that they could decide future conflicts with the dominance of air power...once they overcame those technical shortcomings. The most ardent advocates of this belief became derisively known as the Bomber Mafia.

The Army Air Corps in the 1930s had little sense of purpose and had even less purpose for bombers. Leon Johnson, a future head of the Strategic Air Command, attested to that fact:

We flew around...and did our training because that's what you did in peacetime. I know that we didn't have a sense of purpose at that time. We didn't see anything on the horizon; we weren't worried about anything. We were just worried about getting enough airplanes to fly, and we were worried about getting our flying done.[10]

Although tested only under impossible-to-duplicate optimal conditions, daylight precision bombardment became the primary airpower strategy of the USAAF in its planning for future conflicts. The core members of the Bomber Mafia produced two airpower war plans: the hastily written Air War Plans Division Document One (AWPD-1), and its more deliberate successor, Air War Plans Division Document 42 (AWPD-42) which guided the wartime expansion and deployment of the USAAF.

Henry H. Arnold (1886–1950)

Henry Harley Arnold[*] graduated from West Point in 1907 with a commission as a second lieutenant of infantry. While wanting the

---

[*] Arnold's most widely *known* nickname, "Hap," was short for "Happy," but he preferred it only by intimates: its use by strangers irritated him.

cavalry, Arnold accepted a commission in the 29th Infantry Regiment,* then in the Philippines. Finding troop duties distasteful, Arnold volunteered to help another officer map the island of Luzon. When that detail was done, Arnold expressed an interest in the Signal Corps…but got no response. On his way to Fort Jay, New York via Europe, Arnold saw Louis Bleriot flying his monoplane in Paris. Still stuck in the infantry, Arnold applied to the Ordnance Corps because he could get a quick promotion. Then fate intervened, and he applied for transfer to the Signal Corps, and found himself assigned to Simms Station (Huffman Prairie) near Dayton, Ohio for flight training in April 1911 with Orville Wright himself. Arnold became the US Army's second licensed pilot after just under twenty-four hours of instruction.

After a crash that nearly killed him, Arnold argued against the separation of the Aviation Section from the Army's Signal Corps as being premature. On his next tour in the Philippines, Arnold quartered near George C. Marshall, who became a mentor and friend. By the end of WWI, Arnold was a full colonel—the youngest in the Army—and the executive officer of the Air Division of the US Army. Arnold finally reached France on the day of the armistice.

Between 1918 and 1938, Arnold served in a variety of assignments, including attendance at the ACTS and the Army Industrial College. For the sin of speaking out on the behalf of Mitchell at his court-martial, the Army exiled Arnold to Kansas. But he stayed active enough that his many awards—including three Mackay Trophies for aviation achievements—made him noticeable, and in September 1938 President Roosevelt appointed him Chief of the Air Corps with the rank of major general.

The years before Pearl Harbor saw Arnold pushing hard for heavy bombers; the kind Mitchell and the ACTS advocated. During *that* time, Arnold saw the B-17, and the B-24 proposed and accepted for prototyping. Throughout his career, he boosted the development of the Navy's calculating bombsight that the ACTS *insisted* would win wars swiftly and bloodlessly.

By 1944, Arnold was the chief of staff of the USAAF and the lead cheerleader for the product of an October 1938 requirement emerging from a 1935 feasibility study for a 5,000-mile range superbomber issued by Oscar Westover, another bomber advocate from the ACTS. When the

---

* At this time, West Point graduates did not *have* to accept commissions at all. Rather, they were offered *jobs*: rank below general belonged to the units, not the individuals.

Navy wrote off the Philippines in the first stages of its early war plans against Japan, Arnold began thinking seriously considering Boeing's submission for the superbomber, which he approved for prototyping in 1941. He had pitched what would become the B-29 to Congress as a budgetary plan* for the defense of the Panama Canal, Alaska and Hawaii.

However, since blockade—and *not* invasion—was the Navy's prewar plan for Japan *when*, and not *if*, that war started, long-range bombardment of the Japanese archipelago would be the perfect laboratory for Mitchell's disciples to prove that their bombers *were* the weapons of destiny.

Hayward S. Hansell (1903-1988)

Hayward Shepard Hansell Jr. was born of genteel Southern stock, the son of an Army surgeon and a Georgia socialite. Earning a degree in mechanical engineering from the Georgia School of Technology in 1924, Hansell went to work as a boilermaker's apprentice in California before he joined the Army Air Corps.

Small in stature and given to histrionics, the AAC thought of Hansell as their unofficial poet-laureate, assigned to pursuit squadrons for much of his early career. The Army sent Hansell to the ACTS in 1934, after which he became an unflinching and unapologetic member of the Bomber Mafia. Observing the London Blitz in 1940, Hansell developed a horror of area and night bombing, and insisted that only precision daylight bombing was humane enough for the Americans to use.

Hansell was one author of AWPD-1, a document that spoke of daylight precision bombardment as if it were a settled fact that the United States would only use such humane and efficient ways of war. As a combat commander in Europe from 1942, Hansell learned that not only did ground-based air defense radar make the bombers much more vulnerable, the fighters and the antiaircraft artillery were far more effective than ACTS had calculated. He also learned that industries, both German and British, were a great deal more resilient than ACTS believed. In May 1943, Hansell flew his last combat mission in Europe, and Arnold transferred him to the planning staff at the Army War

---

* An internal euphemism for a bait-and-switch scheme, where the intended purpose for a weapon system was different from the one declared to obtain funding. Before WWII, both the British and Japanese armed forces also performed the same sleight-of-hand.

College, and then to Washington to plan the bomber offensive in the Pacific using Boeing's B-29.

Curtis E. LeMay (1906-1990)

Curtis Emerson LeMay was the son of an itinerant worker and a housecleaner who had to work or fight for everything he ever got. He worked his way through Ohio State University while taking classes in civil engineering and accepted a commission as a second lieutenant in the Army Air Corps Reserve in 1929 so he could learn to fly on the cheap. After receiving a regular commission in the Air Corps in 1930, he completed his degree after finishing flight training in 1932. First trained as a fighter pilot, LeMay transferred to bombers in 1937 and was one of the first Army officers to receive navigator training. He was the lead navigator when Caleb Haynes water-bombed USS *Utah* off the California coast in 1938—despite the Navy giving the Army the wrong coordinates—during a navigation/coastal defense challenge that the Navy lost.

In 1942, cigar-chomping[*] LeMay helped Hansell develop the combat box defensive bomber formation, and eventually took command of the Eighth Air Force's 3rd Air Division, where he earned a reputation as one of the most successful, aggressive, and innovative air commanders in Europe, and as an "operator," a guy who did whatever it took to get results.

Lauris Norstad (1907-1988)

Lauris Norstad graduated from West Point in 1930 and branched to the cavalry. Like many horse soldiers with ambition in the '30s, Norstad transferred to the Air Corps and by 1933 was commanding a fighter group in Hawaii. In keeping with his meteoric rise in the ranks, Norstad was the officer in charge of the 9th Bombardment Group's Navigation School in 1939, and was Assistant Chief of Staff for Intelligence, Air Force Command Headquarters near the District of Columbia when Japan attacked Pearl Harbor. Arnold noticed Norstad's talents and drive, appointing him to his Advisory Council in February 1942. In August of the same year, Arnold named Norstad Assistant Chief of Staff for Operations of the Twelfth Air Force, and sent him to Europe to help plan the invasion of North Africa that November. There, he would rise to

---

[*] LeMay suffered from Bell's Palsy from an infection he got in Europe, and clenched cigars and pipes to keep his face from drooping.

Director of Operations for the Mediterranean Allied Air Forces (MAAF). In 1943, at age 36, he became the youngest three-star general in the Army Air Forces.

The Twentieth Air Force

At the Casablanca conference in 1943, President Roosevelt proposed to send bombers to China to bomb Japan and, as a "practical" matter, help the Chinese as well. While China's Chiang Kai-shek was happy to accept American largess and Arnold was ecstatic to get presidential support, Army Chief of Staff George C. Marshall pointed out the many impracticalities. But, like many such presidential proposals, the plans went ahead. The proposed bases were in the area around Chengtu in south-central China's Sichuan province, north of where Claire Chennault's Fourteenth Air Force was based. But such plans were impractical in 1943. Churchill and Roosevelt also asked Stalin if they could station bombers in the Soviet Pacific maritime provinces. Stalin denied their request, citing the Soviet Union's neutrality to Japan.[11]

The Twentieth Air Force came into being on 4 April 1944 in Washington, DC. The original plan stated that this new numbered air force would have three bomber commands—XX in India and China, XXI in the Mariana Islands once *they* were in American hands, and XXII in the Philippines before the invasion of Japan.[*] Each Bomber Command comprised bombardment *wings* of several bombardment *groups*, each of 30-50 aircraft. Arnold would equip all three bomber commands *only* with B-29s.

On 10 April, the Twentieth Air Force became an agency unto itself. The Joint Chiefs of Staff directed theater commanders to build whatever Arnold needed to accommodate the B-29s and its supporting aircraft. This gave Arnold a *carte blanche* to dictate *where* he wanted *his* bombers to go and *when*. But while *he* commanded Twentieth Air Force, the Army Chief of Staff, in consultation with the Joint Chiefs, commanded *him*. While Arnold *looked* independent, he was really subordinate to a national command. Furthermore, he needed the cooperation of the Navy (especially) for carrying his bombs and equipment and for building many of his airfields and bases, and that of his own service's ground troops—especially supply trains and

---

[*] No other numbered Air Force in the USAAF had three Bomber Commands. Indeed, most only had *one*.

construction engineers—to provide for his bombers. The "independence" of Twentieth Air Force and its big bombers was very much an illusion.[12]

In August 1944, Arnold made Hansell the commander of XXI Bomber Command. Many in the AAF had misgivings about giving Hansell this assignment, saying he was an excellent staff officer, but lacked the killer instinct for aggressive combat command.

That same month, Arnold brought Norstad back to Washington as the Twentieth's Chief of Staff. He would become de facto commander when Arnold was too busy (or ill) to run the day-to-day operations. Norstad was at least indirectly responsible for what was to follow.

Strategic versus Tactical Air Power

The desired result of *strategic* air bombardment is to get the enemy to stop using a resource, or deprive them of it—permanently if possible, briefly if necessary. That meant attacking factories, transportation networks and, ultimately, morale (despite denials to the contrary), the last of which seemed to be the most elusive.

While *tactical* air power (control of the battlefield) was irrefutably as valuable as it had been in WWI, *strategic* air power's value (attacking enemy cities and infrastructure) seemed more attritional than immediately decisive, sparking little prewar interest or enthusiasm (despite Mitchell's and Douhet's assertions) outside of ACTS circles and Great Britain's Royal Air Force (RAF) because no one was interested in a prolonged conflict. Air power enthusiasts in Germany, France, Italy, the Soviet Union and Japan were skeptical of the value of long-range bombardment, did not believe it would either shorten or eliminate wars, and thus did not structure *their* air forces for it.

After WWII started, years of British, American and German bombing in Europe and North Africa failed to trigger the panic or degrade the morale of the populous that some prewar theorists had insisted would occur even as soon as the first bomber appeared over an enemy city, or the first bomb fell. There were even fewer indications that aerial bombardment alone would shorten *any* war or negate the need for direct ground combat. But Arnold and Hansell were staunch believers who clung to the elusive promise of the long-range bomber to a point of obsession.*

Morality, Legality, and the Air War

Unlike surface warriors, post-WWI air advocates were usually *offense*-minded. This bloody-mindedness rarely sat well with their governments or military organizations, as the very vocal Mitchell discovered the hard way. In addition, no one likes to be called what Mitchell called his bosses and their Congressional masters—uncaring. Even if he *was* right, he was *also* insubordinate, and that was no way to sell a completely new way of thinking to a parsimonious country frightened by the high costs of its first and, until then, its *only* involvement in a European war.

Aside from Mitchell's legal issues, there were questions about the legality of strategic bombardment, and there were voices who wanted to ban airplanes as weapons altogether. As early as 1923, the Hague Convention of Jurists attempted to reign in some of the more outlandish air war ideas by declaring that "bombardment from the air is legitimate only when directed at a military objective." Even President Roosevelt proposed practical disarmament of all "offensive" weapons of war, to include bombers, in 1933. Great Britain, who relied on bombers for their air policing campaigns in the Middle East, met this proposal with flat refusals. In 1938, Winston Churchill spoke of the "blackmailing power" of heavy bombardment. That same year, when the Army Air Corps had about 1,800 aircraft—a third of which were obsolescent—American newsman Walter Lippman, referring to the Sudeten crisis that same year, asked Americans how they would feel if there was an enemy air force based within 100 miles of their major cities, "...believed capable of killing or wounding 35,000 to 40,000 people in a single raid."[13]

Mitchell's defenders and disciples knew his wild images of huge bombers capable of 35,000-foot ceilings and a 5,000-mile range were just the thing for the Pacific theater and the war with Japan, a war that some American planners had regarded as inevitable since the late 19th century. Arnold, who had testified on Mitchell's behalf at his court martial in 1925, was one of those dreamers; so was Hansell. While students and faculty *discussed* Douhet in the ACTS, many of those who attended the school never read his book.[14]

---

* Arnold never believed that the land campaign in Europe was necessary to defeat the Axis, and at least a few of his contemporaries agreed with him.

Bombers and the Pacific Theater

Beginning in the 1920s, the US Navy-led American war planners developed War Plan Orange (for Japan). As they argued over priorities, attack routes, and speed of seizure of islands for naval bases, planners for the USAAF thought in terms of airfields. While the fighter jockeys were expressing *their* opinions in the '20s, the others gave more importance to the planned, but as-yet *non-existent*, four-engine bombers, the only airplanes that could conceivably cover those tremendous distances.

Initially, the Boeing B-17 Flying Fortress, first flown in 1935, was to be the mainstay of the Army's heavy bombardment force everywhere. But with a 2,000-mile range, hitting Japan with a B-17 would require bases either in China or the Philippines. Prewar planners did not envision bases in China, and allocated no resources to build them. They also knew that, given the paltry resources available before 1940, any attempt to hold or reinforce the Philippines following a Japanese onslaught would be futile.[*]

B-17G (National Museum of the USAF)

By early 1945, Allied planners started detailed planning for an invasion campaign of Japan. The invasion would involve millions of

---

[*] The decision to abandon the Philippines at the outset of a war with Japan was made as early as 1910, WWII-era propaganda and half-hearted efforts to reinforce it after Pearl Harbor notwithstanding. In the event, Japan cut off the Philippine archipelago from outside support within days in December 1941.

men, thousands of ships and airplanes, and planners projected it to last at least six months, probably a year. Arnold knew as well as any senior officers of the attritional nature of land-based invasions in the Pacific. For example, the Peleliu campaign, which was only *supposed* to take a week…lasted over three months.[*] He and other senior air commanders regarded any serious plans for an amphibious invasion of the Japanese archipelago as foolhardy. He and a few others assured anyone who would listen that air power alone could defeat Japan.

But no one *was* really listening. Why did anyone believe the Japanese would submit to air power alone? The German Blitz did not frighten the British. Germany hardly responded to the massive raids on Cologne and Hamburg and Berlin except with propaganda about bombing schools and hospitals. If anything, the effect on enemy morale was to stiffen its resolve to continue. So where was this *morale-breaking proof?*

The Dread of Fire

By WWII, the British and American insurance and fire protection industries had been studying *potential* targets for years, if not decades— risk *is* their business, after all. Very early in their studies they were looking seriously at incendiary bombing as an effective tool. The RAF/USAAF Hamburg bombing campaign in July 1943 was an experiment in large-scale incendiary bombing.[†] With Japan's severely fire-prone structures,[‡] planners viewed it as a viable option.[15]

Target planners seek to identify critical nodes which, when destroyed, will collapse an enemy's infrastructure. This had been the goal for the USAAF in Europe and had *finally* been successful late in 1944, culminating with concentrated, large-scale air attacks on Germany's petroleum infrastructure.

The best weapon against large, spread-out targets was incendiaries. Extensive testing of both weapons and target structures beginning in 1942 had produced considerable data on destruction potential by testing firebombs on mockup Japanese cities. An October 1943 report by air

---

[*] 15 September-27 November 1944.

[†] A typical bomb mix consisted of both high explosive (HE) and incendiary bombs. Most targets in Europe were hit with a mix of about 15% incendiaries; the USAAF in Europe followed RAF practice. The RAF used up to 40% incendiaries on selected targets like Dresden and Hamburg.

[‡] The building materials of preference in Japan were wood and paper; fire-resistant materials were used in less than 20% of non-commercial urban construction.

intelligence analysts concluded that a thousand tons of the right incendiaries would set uncontrollable blazes in twenty Japanese cities that could slash industrial production by 30% for six months. Most senior officers in the AAF knew of these tests and reports.[16]

One of the earliest AAF voices to suggest firebombing was Emmet O'Donnell,[*] who commanded the 73rd Bomb Wing of the XXI Bomber Command. As early as June 1944, while was training his flyers in Kansas, he came to realize that visual or radar bombing from the B-29s service ceiling would be woefully inaccurate (see below). When Hansell took charge of XXI Bomber Command, he ordered such training stopped immediately. Returning to high-altitude precision daylight bombing, as Hansell required, put Hansell and O'Donnell at loggerheads when they finally got to the Marianas that fall, and depressed morale among the airmen.[17]

But O'Donnell couldn't let it go. Writing a seven-page letter to Arnold on 8 August 1944, he stated,

The Japanese people have never had to submit to a real attack and on the few occasions recorded where they have suffered natural disaster, such as the Yokohama earthquake and succeeding fire,[†] they did not stand up well as a people to their adversity. [18]

The Committee of Operations Analysts

In late November 1942, Arnold formed a Committee of Operations Analysts (COA). This outfit was one of those so common in Washington early in the war, made up of political, legal, military, industrial, intelligence, engineering, and economic experts whose job it was to figure out how to win the war. *This* one, however, was *chartered* to tell the bosses *not only* that something *wouldn't* work and *why*, but how to fix it so it *would* work, as stated in its founding directive, which read, in part:

(d) To study the strategic possibilities to be adopted when current plans have become impracticable, and to advise the Joint Chiefs of Staff in regard thereto.[19]

The COA had the ear of just about everyone who truly believed that all a bomber had to do was destroy the targets that they designated and the war would end. Initially, their task was to tell Eighth Air Force in

---

[*] Called "Rosie" by intimates, but rarely in public.
[†] Known as the Great Kanto Earthquake, 1 September 1923.

Europe how to defeat the Germans so a land invasion could take place at the earliest opportunity, despite Arnold's certainty that it would not be needed. On 14 December 1942, a Colonel CG Williamson, a member of the COA from the USAAF Directorate of Bombardment, was quoted as saying, "…the question as to how many bombs were required to inflict a certain amount of damage [is] a simple mathematical proposition."[20]

Despite this early naivete, the committee, over time, realized that bombing accuracy was not as simple as just dropping bombs as the bombardiers saw them. Accuracy suffered with altitude, wind, temperature and many other imponderables, with enemy action making it just that much worse. By late 1943, their "simple mathematical proposition" was *not* so simple after all.[21]

But, the COA told the bombers in Europe *what* targets to hit, not necessarily *how* to hit them. In November 1944, they shifted their focus to the Pacific. As they had in Europe, the COA recommended bombing steel plants, aircraft factories, radio factories and the like, just like Europe. Regrettably, Japan was a *much* different target than Europe.[22]

The Tempestuous Superforts and their Target

...The situation as I found it was a disgrace to the Army Air Forces...

Henry H. Arnold

The Planes...

When Boeing began making the prototype in 1941, the B-29 project was a substantial risk, as no one had ever built such a plane before, and even the designers didn't trust it. It was a "three-billion-dollar gamble," a moniker given the B-29[*] by the USAAF project development officer for the airplane, Kenneth Wolfe. The first prototype flew in September 1942. Based on those two prototypes, even after one crashed and burned, Arnold ordered *two hundred more.*[23]

---

[*] For comparison, the Manhattan Engineer District built the first atomic weapons for a mere $1.5 billion.

Boeing B-29 Superfortress (Wikimedia)

Boeing equipped the B-29 with four completely untried engines based on the widely used Twin Cyclone, Wright's R-2600 14-cylinder radial, adding two more cylinders per row for additional displacement. The result was the R-3350 Duplex Cyclone twin-row, supercharged, air-cooled, 2,200 hp radial with 18 cylinders with cast steel heads, giving a displacement of 3,350 cubic inches (hence the name). Wright mounted the cylinders on a three-piece forged aluminum crankcase (later made of steel). Each row of cylinders had its own General Electric B-11 supercharger with magnesium cases to reduce weight. Boeing mounted one supercharger on each side of the engine nacelles, which themselves were as big as a P-47 fighter fuselage. Each engine drove a four-blade Hamilton Standard/Hydromatic constant-speed, full feathering propeller, then the largest aircraft prop in the world.[24]

For airborne defenses, the B-29 sported an optically aimed, remote-controlled gun control operated by a TV sighting system. It could link two or more turrets to a single gunner (though the Superfortress carried four), and used analog computers to compensate for airspeed, speed and direction of an attacker, altitude, gravity, temperature and humidity, and the ballistics of the projectile—0.50 inch or 20 mm. It also made corrections for the parallax between the sighting station and the remotely located guns. The idea was for the system's four turrets, two blisters and tail guns to concentrate a broadside of fire on attackers approaching from any angle. Despite the system's sophistication, gunner training was always too brief, even for the elite B-29 gunner teams, commanded by their own officer. Some Superfortress air gunners had never even flown in an airplane before they reported to their operational units.[25]

Based on the experience of aircrews over Europe, Boeing pressurized the B-29 fuselage to the air pressure equivalent of an 8,000-foot altitude when the aircraft was at 30,000 feet. Far from being a luxury, this simply allowed for a central heating system to keep the crew warmer and allow them to breathe without oxygen masks at operational altitudes, thus making them better able to endure far longer flights. The B-29 wing sported flaps equivalent to 1/5 of the area of the entire wing. Each of the two wing spars were the longest and heaviest Duralumin extrusions ever used in a production aircraft. The Superforts carried their fuel in fourteen outer-wing, eight inner-wing, four center-wing and four removable bomb bay tanks, giving a maximum capacity of 9,438 gallons. The designers thought that a fully loaded B-29 would have a *theoretical* operational range of about 6,000 miles; at 300 miles an hour; about twenty hours of flight.

"Operational range" is the *practical* distance a combat plane can fly a round-trip mission with a full fuel, ammunition/bomb/cargo and crew load, providing a 10% safety margin *plus* a further 20% allowance for maneuver, formation-forming and evasion. Operational range depended a great deal on payload and altitude, wind and weather, and scores of other unknowables. Only long experience can reveal the *true* out-and-back operational range of a combat plane, which, historically, is invariably *much* lower than the manufacturer's figure: with the Superfortress, about *half*.

The rush to production of an engineering prototype caused many of the problems. Normally, years of testing and modification on twenty or more pre-production airplanes should have resolved these issues in an orderly fashion after a few months or years. But Arnold imposed a sense of urgency on Boeing that allowed no time to work the bugs out of the most advanced airplane of its time. Although a magnificent, fast, high-flying, and long-ranging airplane, the B-29 suffered from myriad technical issues that defied not only belief but understanding, starting at the *drawing board* and the *factory*....

The Problems...

The B-29 went through tens of *thousands* of design changes during its production-line development, hundreds of them implemented in the field—over *six thousand* in the engines alone. Boeing and the AAF practically rebuilt every airplane and engine *after* they left the factory to incorporate the many improvements and updates that developed *during* production and even during short shuttle flights. Indeed, Boeing built another plant in Wichita, Kansas solely for updating new B-29s built in Renton, Washington, to bring them into compliance with up-to-the-minute specifications. Yet *another* round of modifications was usually needed by the time the aircraft reached their units…and *after*. Thus, every Superfortress reaching their units before mid-1945, when the changes slowed down, was practically hand-built, each one the product of a *B-29 Special Project* that started in 1943. Boeing made the world's most advanced aircraft on the factory floors from Washington to Georgia, and *re*made them from Kansas to Karachi.[26]

Problems with engine overheating were legendary. One aircrewman claimed to have timed an engine fire burning through a wing at just over four minutes, which was one reason Wright replaced the aluminum crankcase with steel. The engineers and crews overcame them with a combination of redesign, field modifications and pilot experience. Pilots and experts at Boeing, Dodge,[*] and Wright believed that the planes' engines didn't warm up properly because they were being flown up to formation and assembly altitudes too quickly, leading to engine fires. Pilots learned to use as much of the runway as possible and build up speed to control engine heating before slowly climbing to altitude.[27]

Addressing these issues gave La Verne G. Saunders of the Second Air Force[†] an opportunity rarely experienced by *any* trainer: the chance to train mechanics *within* that blizzard of engineering changes issued between 10 March and 15 April 1944 at the Wichita plant—an ordeal that came to be known as the "Battle of Kansas." Those mechanics scattered to the other Kansas fields at Smoky Hill, Pratt, Great Bend and Walker, where the Second Air Force trained the first B-29 maintenance crews. Some believe that the B-29's maintainers knew more about their airplanes than did their builders.[28]

Learning to repair the aircraft was one thing; learning to *fly* it was another. Although overall the B-29 was an easy plane to learn to fly, many of the early aircrews were veterans of the B-17 and the B-24, and for them, the learning curve was stark. Constant monitoring of engine temperatures was crucial, especially for the first few minutes. Because of its clean lines, the B-29 slid through the air smoothly which resulted in a time lag between control input and the aircraft's response. This meant the pilot had to plan for changes in speed and attitude. In the

---

[*] This division of Chrysler built nearly 19,000 R-3350-series engines during WWII.
[†] Second Air Force was and *is* responsible for training in the Air Force.

Flying Fortress and the Liberator, cutting power made a difference *now*; in the Superfortress, *later*. Trimming a flap on a Fortress changed the attitude *now*; on a Superfort, in *perhaps* another few seconds. Some pilots called banked turns in a Superfort "flying on ice." Some B-29s responded as though rudder changes were mere suggestions to be followed when it got around to it. Other Superforts were easier to control by varying engine speed or prop pitch. These behaviors spilled over into formation flying, making the pilots provide bigger safety margins between aircraft, which necessarily spread out the formations.[29]

American pilots and planners thought the biggest combat vulnerability to any bomber was enemy interceptors—and the fighter pilots tried to make sure the bomber crews *believed* it.[*] In response, the combat box optimized the defensive gun power of the heavy bomber in Europe. These formations created interlocking fields of fire to ward off attacking fighters. In the B-29, that bigger space needed for control created special problems for tactical defensive planning and, at operational altitudes, made for a formation that could stretch up to 15,000 feet in altitude. For this reason and several others, the B-29s rarely used any formation resembling the combat box.

Japanese intelligence, by some means that are, to this day, still unclear, gathered a lot of information about the B-29 even before its first flight. They could, for example, accurately predict both its service ceiling and its range before they ever laid eyes on one. However, Japan's poverty did not allow her to exploit that information to provide an adequate response.[†30]

Hitting the Targets

Accuracy and *precision* are two related measures of observational error. *Accuracy* is how close measurements are to the requirements. *Precision* is how close measurements are *to each other*. In machine tools and in manufacturing, *precision* is related to *repeatability*. Getting bombs on the target is *accuracy*; getting them close to each other is *precision*. Precision bombardment was illusive because, despite years of experience, factors beyond their ability to control or compensate for with free-falling bombs hampered their accuracy—the *first* requirement.

---

[*] The bigger threat—statistically and psychologically—was anti-aircraft artillery.
[†] As one example of Japan's dire financial situation, in 1922, the IJA was forced to deactivate two infantry divisions to be able to *afford* developing a machine gun, a tank, and a new field radio system.

Three critical issues needed to be addressed to achieve *any* accuracy, let alone precision, in horizontal bombing.

- Getting airplanes *over* the targets.

- Putting bombs *on* the targets.

- Making bombs *effective against* the targets.

The problem in the illustration below addresses just a few of the issues.

Problem 3 (15pts) Targeting:

A bomber flies horizontally with a speed V and at a height h. Ignore the air friction and assume there is no wind. The acceleration of gravity is g. (Express your answer in terms of V, h, and g.)

(a) How long does it take for the bomb to reach the ground?

(b) To bomb a target, how far away from the target should the bomber release the bomb? (ie Find the distance D in the figure below.)

(c) What is the speed of the bomb just before it hits the target?

(d) What is the location of the airplane when the bomb strikes the target.

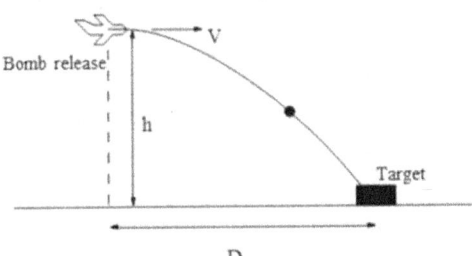

Solution:

(a) The bomb free fell for a distance h. From $h = \frac{1}{2}gt^2$, we find that it takes $t = \sqrt{\frac{2h}{g}}$ for the bomb to reach the target.

(b) The horizontal velocity of the bomb is always V. Thus $D = Vt = V\sqrt{\frac{2h}{g}}$.

(c) The vertical velocity of the bomb before striking the target is $V_{vert} = tg = \sqrt{2hg}$. The speed of the bomb before striking the target is $V_{v\,ert}^2 + V^2 = 2gh + V^2$.

(d) Since the bomb and the airplane have the same horizontal velocity, when bomb striks target, the airplane is right above the target.

Horizontal Bombing Problem (US Navy manual)

During WWI, the US Navy was more interested than the Army in level/horizontal bombing. To address the many issues illustrated above, the Navy contracted for the Sperry/Norden calculating bombsight to achieve a promised "magic 1,000-foot circle" of bomb placement. It compensated for the factors shown above, in part by linking into an automatic flight control (AFC) system—autopilot—to fly the plane straight-and-level while aiming the bombs.

Although both services were to use the device, its designer/manufacturer, the Dutch-born eccentric genius Carl Norden, didn't *like* the Army. He thought its officers were uncouth and ungentlemanly, and thus allocated most of his products and attention to the Navy; he communicated with them only grudgingly. It got so bad that at one point, Army officers had to persuade the Navy to suggest its design changes to Norden, just to make progress. Fortunately, Norden himself had very little to do with the AFC system or its integration with the bombsight that bore his name.*

The Norden compensated for some drift and crosswinds, achieving results that bordered on the miraculous...in ideal *test* conditions...*once*. But the Norden could not address issues of cloud cover, masking smoke, or target camouflage, among many others. The *very* best results the Norden managed in Europe under ideal conditions—which were rare— were about 75% of bombs inside the "magic" circle; the *average* was less than 10%.

There were attempts to supplant the Norden with radar on the B-29. The AN/APQ-13 ground scanning radar system, an improvement over the British H2X, was supposed to enable bombardiers and navigators to see through clouds, smoke and other atmospheric obstructions and make bombing and navigation more accurate. However, the system was finicky and prone to failure. It also required additional training—that most operators did not *get*—to tell the difference between buildings, bridges, roads or other manmade objects, and to distinguish them from ground clutter.† Even identifying forests and mountains by radar was a challenge. The best that *most* radar operators, navigators and bombardiers could do with the B-29s radar was to distinguish between water and land.

---

* The Navy lost interest in horizontal bombing in the '20s, and used bombsights only on seaplanes, technically *patrol bombers*, by the time the Norden was practical. But due to service rivalry, they jealously controlled allocation of Nordens until well into 1943.
† Unwanted electronic echoes from the surface

Japan: A Unique Target

Japan's absolute best defense against air attack was its unique weather patterns. While *appearing* to enjoy a temperate environment, Japan was, from a bomber's viewpoint, a nightmare. Lovely, wispy clouds so common over Japan made target identification from 25,000 feet nearly impossible at least half the time. Screaming jetstreams* deflected bombs as much as 55 degrees off their intended bomb line: the Norden bombsight could only accommodate 45 degrees. Just as bad, the jetstreams trigger severe storms that pile up along the edges of the Japanese coast suddenly and with annoying frequency.[31]

Japan in WWII was a hive of industry but, unlike Europe, that industry came out of Japan's un-rationalized, decentralized industrial development. Because Japan lacked large commercial rivers for transport and had few double-track railroads, the term "industrial target" in Japan placed it within a coastal belt 600 miles long and only twenty miles wide running from Tokyo along the southern coast of Honshu island and into northern Kyushu. This belt contained many factories, but also small machine shops, sub-assembly shops, even individual machine tools operated in private homes and small buildings scattered throughout the cities and towns in the area. *Non*-factory manufacturing amounted to about 15% of all industrial production in Japan. Such a broad industrial dispersion, particularly when combined with Japan's more primitive infrastructure, made the COA's critical-node bombing strategies impractical to the point of being useless. In Japan, critical nodes were not key target *points*, but entire urban areas.

It was difficult for the Americans to say how many people lived in urban areas. The Americans prepared a study called "Estimated Population and Labor Force 1944 for Selected Japanese Cities" which put Tokyo's population at 7.3 million (a little less than 10% of Japan's prewar population), with 3.5 million (52%) considered capable of participating in the labor force. Compared to prewar and postwar studies this was a fairly accurate count, but it assumed only adult participation, failing to note that children as young as twelve were pressed into factory work.[32]

But the Japanese, themselves, weren't much better at determining their urban populations. War mobilizations, food shortages, urban demolitions for firebreaks (see below), and evacuations made every list

---

* First observed in the 18th Century, WWII was the first time pilots had to fly in them.

and census better suited for kindling than planning. Though city authorities charged their ward authorities and fire wardens with knowing who was where, people living in wood and paper houses with holes covered by planks or sheet iron for protection (see below) had a tendency to move at will, often for food and water but just as often to care for sick relations, to be closer to work, or for better accommodations.

Protecting the Japanese People

In 1941, on the eve of the Pearl Harbor attack, Prime/War minister Tojo Hideki pronounced:

Preparations for homeland air defense must not interfere with the operations of our armed forces overseas.[33]

This pronunciation set the tone for civilian defense for the rest of the war. Japanese civil defense was a hodgepodge of makeshift and compromise, mired in samurai ideology and hubris. The most compelling defenses Japan could put up against air raids was wishful thinking. Most people in Japan believed their oceans protected them from air attack, and Japanese military leaders paid little attention to the rapid growth of American and European air power. However, the Dolittle Raid plus the news of the B-29's expected potential gave them pause, providing the motivation to improve air defenses as much as their meagre resources would allow.[34]

As late as January 1944, Japan seemed to act on the assumption that it remained impossible for the United States to capture the Marianas, forcing them to operate their B-29s out of China or even Alaska, both of which would *surely* prove logistically impractical. Even in late summer 1944, when the high command *knew* the B-29s were coming *from* the Marianas—and soon—Japanese authorities denied the danger posed by aerial bombardment of the Home Islands. The samurai mentality, which valued attack over defense, was a contributing factor to this wrong-headedness because air and civil defense were still *defense* and therefore had a lower priority. Their philosophy of "hit hard and often; no defense needed" had not *completely* worn out its perceived validity by 1944.

Cities were *supposed* to have been preparing for air raids since 1929, but there was little evidence of it before the war, even if the *law* was there. The Air Defense Law passed in 1937, and the Greater Japan Air Defense Association began training and propaganda programs in 1939. Because large-scale fires had been a way of life for Japanese cities for as long as anyone could remember, most city dwellers and policy makers were extremely pragmatic about both fire protection and prevention.

There was little *perceived* need for extensive, professional fire protective services in Japan. Nonetheless, a fire department of fewer than 9,000 brave souls served all of Tokyo, one of the largest cities in the world, which covered an area of nearly 213 square miles—ten times the size of Manhattan. Their mostly outdated equipment amounted to barely 1,100 pieces—including Japan's few working* aerial ladders—with fuel allocation limited to two hours per fire truck. Worse still, at least 20% of this meager inventory was out of service for lack of spare parts by 1945. Finally, Japan's fire departments were under the control of the police departments which, themselves, had little or no training in fire protection, and had even less inclination to learn. In March 1945, the Volunteer Fighting Corps,[†] themselves having received only rudimentary fire training, joined the firemen on the fire lines.[35]

Even though Japanese consular officials in Europe described the destruction of Cologne, Hamburg, Dresden, Coventry, and Birmingham in some detail, there was little heed paid to them. While the authorities certainly tried to make the populous *believe* that precautions against air attacks were unnecessary, they weren't so foolish as *not* to prepare for the possibility, conducting regular fire drills. All Japanese citizens carried heavy cotton flame hoods wherever they went, not unlike the omnipresent gas masks of Europe. Each Japanese household was required by law to maintain buckets of water, and every business and public building had at least one water hose and a cistern. And because there were few communal shelters, each household also had to dig trenches to hide in during bombing raids. Authorities tested each of Tokyo's 98 air raid sirens at least once a month.

Though drills and precautions increased as the Pacific war brought the American bomber bases ever closer, the measures taken were crude and useful *only* against small fires. As a partial response to the warnings from Europe, urban authorities created block-wide[‡] firebreaks between major landmarks, usually rivers, beginning in November 1943, destroying nearly a quarter million homes, shops and factories in Tokyo alone, and completely clearing the area around civic buildings, especially

---

* Sources differ as to how much fire equipment of what types were in Tokyo, and how much of it was operational.
† Late war armed civil defense units also called Civil Guards in some texts.
‡ The dimensions were inconsistent—central urban planning not having reached Japan until the 1920s—but usually the firebreaks were about a hundred yards wide.

shrines and Imperial structures. People displaced from the cleared areas often just added to the congestion of the already overcrowded wards.

Civil authorities proposed evacuations of children and non-combatants from urban areas in December 1943, but these gained little traction until the first raids on the Home Islands in June 1944. By September 1944, 411,000 children and their teachers had evacuated Tokyo, Kobe, Osaka, and Nagoya to the countryside. Haphazard planning of evacuations for adults resulted in uneven implementation. Though these measures seemed active, the best protection against air attacks for those who still lived in urban areas would have been bomb shelters requiring scarce concrete and steel building materials. The civil authorities built *some* shelters, but most civilians had to rely on plank- or sheet-steel-covered trenches wherever they could dig them.[36]

Japan's leaders knew they were vulnerable to air attack, but the IJA's insistence that mass air raids on the Home Islands simply were not possible masked the extent of that vulnerability. The protection measures that *were* taken, however well-meaning, were ultimately all but useless against what was to come, and served only to bolster a false sense of security that was based on samurai assurances that large-scale air raids simply could not happen.[37]

Japan's Ground-Based Air Defenses

There was no air defense worthy of the name in Japan before the Dolittle Raid. Ground air defense measures suffered from the same disadvantages as did the rest of Japan's armed forces: poverty and scarce resources slowed technical developments of all kinds, besides the samurai's disdain for anything defensive.

The IJA 1st Antiaircraft Division protected Tokyo, but "division" sounds more impressive than it actually *was*. The unit comprised eight antiaircraft gun regiments and a searchlight regiment. Each gun regiment comprised two firing batteries and two searchlight batteries. With spares and transfers to and from other areas, there were up to 93 guns in *all* the firing batteries of the division at the *most*, of which *perhaps* 80 were operational at any one time.[*] Including an unclear number of independent AA units assigned to specific factories, the Imperial Palace and harbor facilities, over 400 heavy guns and an unknown number of

---

[*] The structure of the IJA artillery batteries varied widely, but usually consisted of 4-6 guns of a single caliber with from zero to three spares at battery and regiment level; land-based IJN gun batteries likely followed IJA protocol.

smaller guns ringed Tokyo. By October 1944, there were at least 1,200 IJA antiaircraft guns in the Home Islands altogether. Any ships in or near the harbors and coasts joined the small number of IJA AAA units then in service. Naval antiaircraft gunfire had the virtue of mobility, but at the cost of a lack of radar or even searchlight-based fire control.[38]

The mainstay of the IJA's antiaircraft artillery was the Type-88 75 mm antiaircraft gun first fielded in 1928, considered obsolete by 1941, and withdrawn from overseas service. The IJA deployed them in the Home Islands before the Dolittle Raid, but did not expect to *need* them, giving them small crews and little ammunition for practice. A new Type-99 88 mm gun that first appeared in 1942 was a reverse-engineered copy of a German SK C/30 88 mm pedestal-mounted weapon captured in China. Neither the Type-88 nor the Type-99 could *effectively* reach the B-29s cruising altitude. The Type-3 120 mm gun, fielded in 1943, *could*. However, Japan only manufactured 120 of them and sent some to Palembang in the Netherlands East Indies to protect the oil fields. The IJN used some ship and ground-mounted Type-88s and the Type-10 120 mm gun, first fielded in 1921, which had a longer range. All these weapons sometimes had to limit their firing to as few as 20 rounds per day because of chronic ammunition shortages.[39]

More sore points to Japan's air defense were early warning and fire control. While the West used radar extensively for air defense, Japan simply couldn't afford to do the basic research, and lacked the resources to even *copy* foreign models effectively. Duplication of effort by the IJA and the IJN affected everything. Each, independently, developed its own land-based early-warning radars, but neither cooperated with the other, delaying deployment of operational systems. Each service duplicated the other's mistakes, and independently discovered the same problems.[40]

The Type B, or Ta-Chi 18, was Japan's best surface-search radar set, with a maximum range of just under 200 miles—less than an hour out for the B-29s. Radar stations in the Volcano Islands could provide at least some early warning for the Home Islands from attacks from the Marianas until the Americans captured them in March 1945. Japanese early-warning radar lacked information on attack altitudes. When the Americans varied approach altitudes, Japanese interceptors were thousands of feet above or below them. Making matters worse, the fragmented radar development program led the Japanese to use a narrow band of frequencies for both gun-laying and searchlight-control radars. American jammers quickly found those frequencies and jammed them, leaving them blind. The best fire control radar was the Ta-Chi 31, which used some German innovations from a smuggled Wurzburg set. It suffered from poor communications, lack of spare parts, and small numbers. Finding and ranging the target, ultimately, depended on optically aimed searchlights. Each service deployed an independent early-warning network, often with IJA and IJN stations located a few miles apart. Japan's clan-based infighting, poverty and poor/late industrial development made it vulnerable to America's air power and technology.[41]

Air raid warning ultimately relied on naval, military and civilian surface observers, and on radio interception. US submarines severely mauled the IJN's picket squadrons starting in 1944. Japan's best, most reliable and longest-lasting asset was three air-signals intelligence units that listened to American radio traffic. Since the B-29 crews frequently ignored communications security protocols,[*] these units could produce reasonably accurate estimates *when* the bombers might appear…but not *where*. Knowing that a raid was coming was a far cry from being able to intercept it. While it was not possible for the USAAF to fool the Germans with diversionary attacks, it was *possible* but not *easy* to fool the Japanese.[†] Against the Japanese, their chattiness often gave them away. The Japanese also did some communications jamming. We know little of it other than that it *was* effective.[‡][42]

Active Air Defense of Japan

After the Dolittle Raid, both the IJA and IJN assigned more and better fighter units to the Home Islands, but they didn't really expect to need them. To each of the three main industrial areas of Japan—Tokyo, Osaka-Kobe, and the Shimonoseki Strait—the IJA assigned seventy fighters, drawn from two new fighter units and three training units converted to air defense in the summer of 1942. In March 1944, the IJA expanded its brigade guarding Tokyo to a flying division (the 10th) and formed a flying brigade (the 23rd) to protect Nagoya.[43]

Both services operated independent fighter and air defense systems. The IJA assumed primary responsibility for the air defense of the Home Islands, while the IJN deployed its air defense only to defend naval bases,

---

[*] Movies notwithstanding, American flyers were notoriously chatty in WWII, a fact noted by German, British, Italian, Japanese and even Soviet listeners.

[†] The Germans had seen RAF diversions for three years by the time the USAAF began its European bombing campaigns; the USAAF attempts at diversion were poor.

[‡] Japan was the first country to use wireless jamming, starting in the Russo-Japanese War (1904-05).

fleet anchorages, and shipyards. Worse, both services organized their active air defense geographically—there were no integrated operations rooms like those of the RAF and *Luftwaffe*. Ground controllers did not vector fighters to intercept the bombers. Instead, they passed word to their units, which launched aircraft independently and controlled them from the ground. Instead of coordinated attacks by large groups of fighters as seen in Europe, US formations faced multiple attacks by smaller groups. IJA assigned their fighters to air divisions which belonged to field armies, and stayed within that army's geographic boundaries. IJN aircraft remained tied to the immediate area of its assigned fleet. As a result, while there were 300–400 fighters stationed throughout southern Japan, only 70 or so would *normally* meet an attack at one time, though sometimes repeatedly as long as their fuel and ammunition lasted.[44]

A single IJA night-fighter unit existed,[*] the 53rd *Sentai,* coming on line in March 1945 and based at Matsudo, northeast of Tokyo, that flew Kawasaki KI-45 *Toryu* (dragon-slayer)[†] twin-engine fighters. IJA also flew the Nakajima J1N1 *Gekko* (moonlight)[‡] twin-engine fighter (used in much of the war for reconnaissance) in the 10th Air Division trained but barely equipped for night interception. Neither fighter had effective radar because the FD-2 sets, based on the purloined Wurzburg, were not worth the additional weight.[45]

Nakajima designed the KI-44 *Shoki* (a contraction for *Zhong Kui*, a mythological demon vanquisher)[§] for the IJA after Japanese intelligence confirmed the development of the B-29. Going into limited production as the first Superfortress prototype flew, it was Japan's first attempt at a high-altitude heavy interceptor. However, it was ill-equipped for the role. Of the different versions produced, only one had sufficient firepower—four 40 mm cannon—to destroy a B-29 optimally, but *only* if the B-29s were low enough for a good pass; most KI-44 versions lacked a powerful enough engine to intercept B-29s at their optimal height. Even so, the "most powerful" version of the KI-44 carried only ten rounds per gun, making a kill more likely by pure luck. The KI-44 did much of its killing by ramming, with the pilot either bailing out or wrestling his damaged aircraft to the ground.[**] Later interceptors like the

---

[*] A second unidentified night fighter unit has been alleged.
[†] Called "Nick" by the Allies.
[‡] Called "Irving" by the Allies.
[§] Called "Tojo" by the Allies.
[**] Both happened.

KI-87 and the KI-94, both designed to intercept the B-29s, were too little and too late.[46]

Nakajima KI-44 *Shoki*/Tojo (Wikimedia)

The fighter pilots protecting Japan learned a great deal from their German allies and their comrades in the Southwest Pacific about fighting American bombers. Superfortress crews reported Japanese fighters attacked mostly from the front and above, mimicking how the Japanese had learned to attack other heavy bombers. Individual fighters usually had limited success, but a swarm of fighters attacking from the front could often confuse the gunners and cripple or destroy a Superfort. Stern attacks were rare, as few Japanese fighters could catch up with a B-29.[47]

Air-to-ground communications were as poor as the radars. Controllers sent night fighters in the expected direction of incoming bombers and then expected them to use onboard radar to find targets, providing they *had* them and they *worked*. All Japanese aircraft and radars lacked Identification-Friend-or-Foe (IFF) beacons or detectors, leaving their operators unable to tell if a radar contact was a target before making visual contact.[48]

As 1944 went on, some Japanese were uneasy about all the assurances of the military. Had they not promised Saipan was inviolate? Had not Tojo resigned in shame for its loss? Were not American bombers reaching Japan already? As for their defenders and first-responders, most only hoped that the rumors of mass air raids were false, because the Americans had done nothing like it before…

Before The Fire Blitz

*A country which cannot defend itself from aerial attack will find its
air bases, its munitions centers, its military depots, its shipyards,
and its great cities subjected to a devastating rain of bombs within a
few hours of the declaration of hostilities.*

P.R.C. Groves

India, China, and XX Bomber Command

On 13 January 1944, Kenneth Wolf, first commander of XX Bomber
Command, and his staff landed in New Delhi, India, to build the
bases there and in China, borrowing construction engineer battalions
from Joseph Stillwell in Burma. Two bases in India were marginally
completed by 2 April 1944, when the first B-29s of the 58th Bomb Wing
arrived, having completed a 11,000 mile flight from Kansas. It would be
September before the four planned Indian bases were ready. This was the
beginning of Operation MATTERHORN, which was to be the first part
of a two-pronged air attack on Japan. While flying to Chengtu on 24
April, the Superforts encountered their first Japanese fighters.

440ᵗʰ Bomb Group B-29 in India (Wikimedia)

The KI-43s* stayed well out of range of the bomber's machine guns and spent some ten minutes studying what postwar Japanese fighter pilots would call "stupendous giants" and "tremendous bulls."

After this unusual aerial once-over, the KI-43s attacked, scoring hits on one B-29 while the Superforts claimed one probable kill. Though otherwise unremarkable, the engagement is notable for two reasons: it dispelled the Americans' concerns about potential disaster resulting from projectiles piercing the B-29's pressurized fuselage envelope,† and it gave the Japanese visual confirmation of the airplane they had been hearing about for years.[49]

The IJA's *Ichi-Go* offensive beginning in April 1944 and lasting until December was an attempt to deny the Superforts bases while also grabbing rice harvests in some of the richest regions in China. Although *Ichi-Go* disrupted some operations briefly, the B-29s still used the Hsinching, Kwangsan. Chengdu and Pengshan bases near the foothills of the Himalayas.

---

* Allied code name Oscar.
† The cabins were depressurized during combat, and overpressurized to compensate for minor damage.

On 6 June 1944, XX Bomber Command launched its first bombing mission from India, Mission #1[*], against targets in Bangkok, Thailand. Of the ninety-eight Superforts launched, only 48 dropped bombs on the target. The resulting damage was negligible because bad weather forced them to use radar for targeting which proved to be even less accurate than visual bombing. On the same mission, it suffered its first losses, when five planes had to ditch. Though an unspectacular beginning, it *was* a beginning.

XX Bomber Command's next target, selected by COA, was the Yawata steel works on Kyushu, Japan. Staging forward to their bases in China, they launched the attack on the evening of 14 June, reaching the blacked-out target at 11:30 at night. Only 47 bombers unloaded on the target, hitting nothing. Twentieth Air Force lost seven aircraft on this first raid on Japan.[50]

The missions that followed out of both India and China were similarly disappointing, and Arnold and the Combined Chiefs were unhappy with both the operations tempo and the bombing results for XX Bomber Command. On 7 July, Arnold replaced Wolfe with his deputy, Saunders, until LeMay reached India on 20 August.[51]

European experience dictated that the bombers should climb to altitudes of 20,000 feet or more to avoid antiaircraft fire and to make intercepting fighters work that much harder. The B-29s initially followed that guidance. But no one told the Japanese fighter pilots just how effective the combat formations were, box or vee, because they pressed home their attacks with suicidal determination…and got some results. While Japanese flak was nowhere near as effective as German or even Italian, it *was* still a grim reaper. Combined, Japanese defensive measures knocked out roughly 2% of the B-29s on every mission—an unacceptable level for the Americans, who at the same time were suffering a *minimum* of 10% operational losses[†] on every mission.

LeMay was supposed to improve the situation, but he soon discovered that only minor improvements were possible on such a makeshift operation, which was shaping up to be a large-scale experiment in expedience on hand-built airfields relying on hand-to-mouth logistics. Operations continued with indifferent results until Mission #21, a large-scale raid on Hankow (Wuhan), China. LeMay sent sixty aircraft to Hankow on 18 December; fifty-five dropped mostly

---

[*] Each mission of Twentieth Air Force was numbered chronologically. Each bomber command kept their own tally.

[†] Mechanical and other failures that caused aircraft loss.

incendiary bombs…and Hankow burned for three days. They destroyed as much as fifty percent of the target area with only a third of the bombs dropped.[52]

Expecting the Superfortresses to bomb Japan as effectively and, more important, as *efficiently* as the USAAF and RAF were bombing Europe,[*] Arnold *expected* the big planes to *easily* reach Japan from Chinese fields as soon as they were operational, flying above the flak and fighter defenses, and taking the war to the enemy's homeland, destroying important strategic targets with their enormous bombloads. But even COA was at a loss with targeting. Given what little definitive information they had about Japanese industry in China and Southeast Asia, LeMay later stated that his best recourse was to choose targets from a world almanac.[53]

However, XX Bomber Command also had to contend with the logistical nightmare to simply *reach* Japan or any *other* targets from China. The B-29's Chinese bases were all at least a thousand air miles on the other side of the Himalayas from their Indian "main" fields. On both sides of the mountains, it only took a couple of hours for the all-too-frequent rain storms to turn the unsurfaced airstrips into mires, grounding everything. Sending *one* B-29 to *any* target from a Chinese field required *ten* B-29-loads of fuel, ammunition, parts and supplies.[†] Even when the bombers reached their targets, the results, save the Hankow mission, were worse than disappointing. After ten thousand B-29 sorties[‡] flown against targets in Japan and East Asia, XX Bomber Command could point out not one mission to suggest that air power alone might defeat Japan. The huge promise of long-range bombardment in the largest and heaviest airplanes to date—with a payload equivalent to the *weight* of a B-17—fell very short in practice.[54]

XX Bomber Command operated out of their crude bases in India and China until March 1945. The XXI Bomber Command bases in the Marianas became operational in October 1944, and were more suitable for the task at hand, which was to bomb the Home Islands until Japan submitted, making amphibious invasions unnecessary.

---

[*] Arnold and nearly everyone else had an exaggerated sense of the Combined Bombing campaign's efficiency.
[†] Infamously, one supply aircraft in twenty carried *only* engineering change documents and parts for the aircraft.
[‡] A sortie is the operational mission of a single aircraft.

The Marianas and XXI Bomber Command

The *first* issue with the deployment of the B-29s to the Pacific was the sheer *size* of the Pacific Theater of Operations. With the Japanese Empire at its greatest extent in mid-1942, there was no airplane in the world that could fly from any Allied-held location on its periphery to *and from* any Home Island target. The prewar plans for blockading and bombing the Home Islands called for bases in the Mariana Islands—Guam, Tinian and Saipan, each 1,500-plus miles from Japan. Saipan and Tinian were within the Japanese Mandates awarded after WWI. She had been colonizing and developing them since she'd moved in, in 1914.[*] Guam, captured from Spain in 1898, was home to US Navy and Marine bases when the Japanese captured it in December 1941.

The Army Air Forces waited for the Navy and the ground troops to first *take* the Marianas. These islands were two thousand air miles from the nearest Allied *anything* at the beginning of the war. By 1943, there were over 30,000 Japanese civilians on Saipan, and almost as many Japanese soldiers and sailors were ready to fight for it. That same year, Tinian had just over 15,000 Japanese civilians with a garrison of 8,500 soldiers and sailors. There were another 22,000 Japanese soldiers and sailors and as many native Chamorros on Guam.

---

[*] Germany had bought them from Spain after the Spanish-American War, but never colonized either island.

Moving a B-29 Wreck, Guam, 1945 (Wikimedia)

Securing the Mariana Islands in the summer of 1944 cost 2,100 American and 85,000 Japanese lives.[*] Before the shooting stopped, the Navy Seabees began building airfields with runways longer and wider than any they had ever built anywhere to accommodate the B-29s. And the Superfortresses, even on a good day, needed every foot of that extra length, and that wasn't always enough. It was horrifyingly common for a heavily loaded bomber to crash off the end of a Marianas runway.

The first XXI Bomber Command B-29s reached Saipan in October 1944. During this period, friendly aircraft had to approach the island following a specific, fixed flight profile to avoid being shot at by island defenders—Japanese intruder attacks were frequent.[55]

Reliable, up-to-date target information on Japan was scarce. On 1 November 1944, just days after the first 3rd Photo Reconnaissance Squadron planes arrived at Saipan, Ralph Steakley flew his photo-recon-configured B-29[†] over Tokyo for 35 minutes at 32,000 feet under ideal conditions—never again seen by XXI Bomber Command—producing

---

[*] This includes nearly all the soldiers and sailors (including Nagumo Chuichi, commander of the Japanese fleet at Pearl Harbor and Midway) and an unclear number of non-combatants.
[†] Designated an F-13A.

excellent photos that planners used for the rest of the war. The Danish minister to Japan, Lars Tillitse, witnessed the extraordinary sight in Tokyo:

It flew very high—about 8000 meters—but in the clear autumn sky, it was plainly visible. It looked like a small silver bird, and it drew a white stripe of smoke over the blue firmament...[56]

Twenty more recon missions over Tokyo failed to produce a single target-planning-usable photograph. Reliable planning maps of *most* of Japan were nearly non-existent for much of the war.

XXI Bomber Command executed its Mission #1 against Japanese submarine pens on Dublon Island in the Truk Atoll on 27 October 1944. Of eighteen airplanes launched, fourteen bombed the target from 26,000 to 27,500 feet. Bombing results were called "poor to fair." Japanese fighters attacked twice, and the bombers reported "inaccurate, intense" flak.[*] The next two missions flown against the same targets yielded the same disappointing results with more airplanes. Mission #4 was against the two airfields on Iwo Jima on 5 November, when 36 Superforts hit the targets with results similar to Truk. Mission #6 went back to Truk on 11 November, but with only a handful of aircraft. Not a particularly auspicious beginning for a plane in which America had invested so much.[57]

And the B-29s waited for the weather over Japan to clear enough for the B-29s to bomb visually from 30,000 feet in daylight. O'Donnell wrote about his frustration in his diary on 21 November:

I have 108 airplanes on hand and not one has done its job...The lack of flexibility has us stymied...I am more than ever convinced that the plan for our initial employment is not good...we have lost an excellent chance to catch them psychologically unprepared...we could have been driving them nuts.[58]

The first B-29 raid on Japan from the Marianas was on 24 November 1944—XXI Bomber Command's Mission #7. O'Donnell in *Dauntless Dotty* led 111 B-29s of the 73rd Wing against Target #357, the Nakajima engine factory at Musashino, a Tokyo suburb, following COA's guidance that placed priorities on engine plants. Because of the weather, only 24 bombers found the primary target, while all but seven of the others found targets of opportunity around Tokyo. They met no Japanese fighters or flak, but the bombs were stuck in several aircraft because of

---

[*] This is how the Twentieth Air Force's Standard for Reporting Antiaircraft fire would term *any* flak: accuracy followed by intensity.

frozen bomb release mechanisms, an issue that would plague future missions. They did no damage to the target (though the day after, some accounts say they *did*), and many Superforts nearly ran out of fuel; some *did* run out and went into the sea. The next three missions against the same target were similar failures, in part because of the weather over Japan and in part due to frozen bomb releases. In his diary, Ugaki Matome, Yamamoto Isoroku's onetime chief of staff recently assigned to the Naval General Staff, described the raids as little more than annoying, but he also wanted to keep people working while they were in the air-raid shelters.[59]

In the first eight missions against Japanese targets, Hansell had lost 23 aircraft, though only seven to enemy action, including collisions. Loss of fuel accounted for some, so did mechanical failures. But eight just…didn't come back for unknown reasons. He also lost a dozen to enemy air raids out of the Volcano Islands,[*] and several more to Japanese holdout guerrillas on Saipan. The crushed coral fill used for building runways, taxiways, hard stands and roads took its toll on both airplane and land vehicle tires—they lasted about three weeks. Coral dust also ate up air cleaners at a breathtaking rate. In addition, Arnold continuously badgered Hansell over the lack of success, aircraft losses, his demands for parts, and criticized every paragraph of every report and communication.

All of that misery and tragic loss might have been tolerable if Hansell's airplanes had been hurting the enemy…but they weren't. Nocturnal attacks out of the Volcano Islands—air raids and commando landings—plagued him with disrupted sleep and destroyed airplanes. As the end of 1944 drew near, Hansell had nothing to show for the effort but bad reports, a quarter of his premier bomb wing becoming casualties, arguments with O'Donnell over tactics that trickled back along the grapevine all the way to Hawaii, poor morale among the men who witnessed the actual issues over Japan, and Arnold's continued bad graces.[60]

What was worse by far for the men was the possibility of being badly injured on a mission. At fourteen hours, missions from the Marianas were twice as long as any mission ever flown in Europe. Sometimes, a crew would bail out a badly injured member over Europe and leave them at the mercy of the enemy. The Germans usually granted clemency to the survivor because they expected—and got—the same from the Allies. But a man hurt over Japan had at least a seven-hour flight

---

[*] Part of the Bonin Islands that include Iwo Jima.

back because, as far as the Americans were concerned, the Japanese showed *their* notion of "clemency" at Bataan, and the Japanese cared not at all for the fates of their own flyers taken prisoner. Aside from that, much of the flight out and back was over water. The Pacific Ocean was merciless to anyone who didn't get to a raft, and even then, they often perished from exposure before they were found and rescued.

But the AAF thought about the on-board wounded, because they trained the radio and radar operators on how to start an intravenous drip plus other somewhat advanced first aid measures, and the squadron surgeons supplied at least two plasma or serum albumin kits on each aircraft.* Without this morale-boosting intervention, because morale starts with hope for the future, the dismal performance of the tempestuous Superforts struggling with weather and engines and everything *else* to do the job might have felt worse than it was.[61]

Regardless of how ineffective it obviously was, Hansell would not back down from his faith in high-altitude precision daylight bombardment. On 3 December, he launched 86 bombers against the Musashino plant once again—Mission #10.† Just 56 Superfortresses reached the target. Post-strike photos revealed that only 1% of the bombs dropped were within 1,000 feet of the target.[62]

On 18 December, Arnold ordered a full-scale incendiary attack on Nagoya just as the lead bombers of the 313th Bomb Wing arrived on the newly built fields on Tinian. Hansell was furious, and came close to actually refusing to order a fire raid, noting that though it *was* in Arnold's name, Norstad, who was long an advocate of both area and incendiary bombing, signed it. The phrase "destroy as much of the city as possible" was the *opposite* of what Hansell and the ACTS had been advocating.[63]

With much regret but following orders, Hansell sent 57 B-29s of the 73rd Wing to hit Nagoya in daylight (Mission #17) on 3 January 1945 from an altitude of 29,000 feet, carrying a little over 139 tons of incendiary bombs and just under twelve tons of fragmentation bombs. This was the home of the Mitsubishi aircraft plant, then the largest airplane assembly works in the world, but the bombloads were badly and ineffectively scattered in the target area. This mission resulted in the loss of five B-29s and caused damage or destruction to just over three acres

---

* Other crewmen were also trained as aidmen by *some* squadron and group surgeons.
† Also called SAN ANTONIO III. XXI Bomber Command *area* raids—as opposed to *precision* strikes—were named using the target's code name and a Roman numeral. Precision target missions only had numbers. Musashino—a large suburb of Tokyo—was given its own name.

(0.005 square miles) of the target. Mission #17 was an unwitting test of a fire raid, where the bombers carried an incendiary-heavy bomb mix, but flew a conventional (high altitude) profile as they would in a conventional mission. However, the Japanese regarded this incendiary-heavy raid as an example of how *ineffective* such attacks would be.[64]

As 1944 closed and 1945 dawned, the big planes and their mostly young crews found themselves embroiled in an engineering, military, and political morass that only a very special commander could address. But Hansell, an excellent theoretician, was *not* that commander.[65]

Senior officers knew that what Hansell kept doing wasn't working, and *couldn't* work. Samuel R. Harris, commanding the 499th Bomb Group in O'Donnell's wing, wrote on 31 December:

> My Group goes in number three [to Nagoya] bombing at 30,000 feet with both frag and fire bombs. So far it is set as a formation attack—bombing visually by squadron—but it may be changed to individual aircraft at night and bombing by radar. I hope the last is true because that type of attack will allow us to use the airplane with some degree of efficiency. We will continue to have mechanical troubles as long as we persist in high altitude formation attacks....About 10,000 feet—individually at night and we can operate.[66]

LeMay Takes Over

In 1943, when they studied the post-strike photos of Hamburg after the six-day Operation GOMORRAH had created the first firestorms and killed 45,000 people, Hansell and LeMay had vastly different reactions. The results appalled Hansell; LeMay saw the necessity. But even after all that destruction, Hamburg's factories production returned to 80% of the pre-attack levels in just five months.[67]

The most important thing the RAF *did* was to prove that the ACTS was wrong: *nodes* weren't the answer, but *cities* might be. But Hansell didn't—or *wouldn't*—see it.

On 4 September 1944, the COA released a study that identified six Japanese (and thirty German) cities that they believed were ripe for area firebombing. The committee thought six Japanese cities—Tokyo, Kawasaki, Yokohama, Nagoya, Osaka and Kobe—to be better area targets than any German targets were, with a greater concentration of workers than the top 25 German targets combined. They also contained a bulk of Japan's manufacturing capacity.[68]

In January 1945, Arnold was in Florida recuperating from his fourth heart attack serious enough to require hospitalization since 1943,[*] and Norstad was the acting commander of both the Twentieth Air Force *and* the USAAF, traveling to the Marianas. Two days after Norstad landed on Saipan, he fired Hansell. Norstad was on Guam when LeMay arrived. As LeMay stepped out of his airplane, Norstad replaced Hansell with LeMay effective 21 January,[†] with Arnold's endorsement and to the apparent relief of many senior officers:

> In [O'Donnell's] office...I was informed that LeMay is replacing Hansel [sic]. Praise be to Allah. There will certainly be changes made around this joint. LeMay is known to be a very good if slightly rough operator—we can take the roughness if there is sense behind it.[69]

In retrospect, Hansell faced an impossible task, trying to implement a strategic bombing campaign with green crews and dangerous, unreliable aircraft against enemy targets more than a thousand miles over water from their bases, and facing weather never previously encountered in air warfare,[‡] using tactics that largely failed elsewhere with different airplanes, based on a theory of warfighting developed in an operational vacuum a decade before.

But "impossible," to LeMay, was a challenge, not an obstacle.

As LeMay plotted a fresh strategy, the Japanese and Americans were destroying the Pearl of the Orient—Manila—leaving behind over 100,000 casualties and thousands of buildings ruined in three months of fighting. On Iwo Jima, after 19 February, Americans and Japanese were killing each other at an astonishing rate, and would eventually leave nearly 30,000 dead between them in just over a month. If LeMay failed, *these* were what the world had to look forward to in an invasion of Japan.

Arnold, whose impatience was legendary, had Norstad deliver an ultimatum to LeMay. He would either bomb Japan successfully with the

---

[*] Some say caused by the heartaches of the B-29 program, but hard evidence is lacking.

[†] Several sources and witnesses offer different dates for this chain of events. LeMay's orders put him in command effective 21 January.

[‡] It has been suggested that Hansell never reconciled himself to the jetstream, a cause of many of his headaches.

Superfortresses, or America would suffer a ground invasion of Japan. That meant George Kenney's conglomerate of air forces under Douglas MacArthur would get control of the world's most expensive and innovative airplane, and LeMay's career would be finished, except for paperwork.

LeMay soon realized he was facing a series of interlocking problems of such great complexity, with so many variables, that only unorthodox solutions could address them all. He later wrote:

I made up my mind to make some major changes in the way we were using the B-29s because it was now clear that we couldn't possibly succeed by basing our strategy on our experience from Europe. That system wasn't working. It was a different war with different weather and a different airplane. It called for a different solution.[70]

To resolve the complex problem of using the B-29 *effectively* against Japan, LeMay would risk not just his career, but the future of the B-29, and indeed, the theory of strategic air power as a *doctrine*.[71]

While Arnold charged LeMay with winning by any means necessary, LeMay told Norstad that, at least for the present, he would maintain Hansell's doctrine until he could address the XXI Bomber Command's low operational efficiency and morale. The men saw an immediate change in attitudes, feeling that they were finally going to war. LeMay instituted schools for *everyone* as he had elsewhere, from the pilots to the mechanics, from the bombardiers and radar operators (especially) to the gunners (who manned machine guns when hunting Japanese holdouts), from the staff officers (many trained as supply men) to the clerks (working shifts as airfield and building maintainers) to the doctors (who became sanitation officers and even vets) to the chaplain's assistants (organized as medical technicians). LeMay scheduled day *and* night training flights and bombing missions, attacking nearby, unoccupied islands, the Japanese bases at Truk…*and* the Volcano Islands, to reduce the pesky and deadly Japanese raids, each mission bombing from successively lower altitudes.

While LeMay kept looking for problems and solutions in the Marianas, he sought answers for the problems in Japan…and he knew the future was flame. For the Americans to bring Japan to her knees, her cities had to burn.[72]

But *morale…throw your oxygen masks away, General LeMay is here to stay!* It became a battle-cry of the B-29 crews in the Marianas.[73]

Planning and Selling a Fire Blitz

The most accurate way to deliver lightweight firebomb clusters was at a low level; the *safest* way was to do it at night. Another practical result of low-level bombing would be airplanes flying *under* the notorious jetstreams. While Japanese searchlights were plentiful, Japanese antiaircraft guns were not known for their accuracy at night.[74]

Getting maximum bomb loads over the target meant eliminating formations, eliminating formation loitering and reducing fuel consumption. Weather, once again, reinforced LeMay's no-formations decision. The weather fronts off the Japanese coast required bomber formations to break up on approach and then reform, a maneuver creating smaller clusters of loitering bombers even more susceptible to fighters and consuming even more fuel. Eliminating the formations diminished both the fighter-interception problem and the formation breakup-reform problem. With little risk of interception, there would be no need for the 0.50 in. machine guns or their ammunition, or the gunners, on each aircraft.

This plan made it look as if the ACTS had never existed. Yet, the B-29's high-altitude daylight bombing capability—*for* which it was *designed*—was the epitome of the ACTS's program for strategic bombardment. LeMay simply decided the ACTS had been completely wrong about *their* approach for bombing Japan.[75]

That said, a planeload of incendiaries was horrifying to any bomber crewman. When a plane loaded with high explosive bombs crashed on landing or takeoff, there was a *fair* chance that at least *some* of the crew would survive, because HE bombs had to be *armed*—electrically or by the downward motion of their descent—*before* they could detonate. But firebombs didn't *need* to be armed; they just had to break open, engulfing plane and crew with searing flames, leaving little chance for anyone or anything to survive. Incendiary bombs also had a nasty reputation for going off inside bomb bays out of sheer bloodymindedness. Few B-29 crews would relish the idea of carrying a full load—about 7 tons—of firebombs.

On 17 February, the Navy attacked the Musashino plant with a massive carrier force, doing more damage to it than Hansell and LeMay had done in a dozen attacks. Suddenly, the USAAF had a competitor in the effort to bring Japan to its knees by air. Unless something drastic happened—and soon—the B-29 would go down in history as a very expensive boondoggle.[76]

For Mission #38 on 25 February, LeMay sent a 172-plane high-level daylight fire raid that dropped 453.7 tons of incendiaries on the commercial and shopping district of Tokyo. Planners considered this the first large-scale test of fire-bombing in Japan. It was the first B-29 mission to Japan that involved three bomber wings, and the first where incendiaries were a greater part of the payload. The results were mixed because the Superfortresses had to aim through a snowstorm, with most resorting to radar. Even so, they burned up a square mile of Tokyo's Kanda and Shitayku[*] wards, destroying nearly 28,000 structures at a cost of five B-29s. These results were better than all the previous high-level attacks combined.[77]

Even as he conducted his tests—having decided to not just *throw out* the book that he himself helped to write, but to *shred and burn it*—LeMay *still* had to sell his scaled-up ideas to his aircrews. On 1 March, a training mission flew to Kito Iwo, a tiny island off Saipan, with a specified bomb-away altitude of fifty (50) feet. Despite some trepidation and not a few questions, the aircrews *carried out* that mission *with* that attack profile. LeMay knew then that the 5,000-8,000 foot bomb-away altitudes he was planning over Tokyo would garner *less* consternation.[78]

Once again, the weather informed LeMay's planning. On 2 March, data from LeMay's weather flights (F-13s), from XX Bomber Command and from the Navy showed that a seven to ten-day window of fairly clear weather over Japan would open up around 9 March. Ever the opportunist, LeMay told his planners to prepare strike plans and field orders for Tokyo, Nagoya, Osaka, Kobe, and Kawasaki at a rate of one mission at least every day, starting on 9 March. He also informed his ordnancemen to uncrate all the firebombs on Saipan, Tinian and Guam, and told the Navy to offload all those in cargo holds afloat.

His planners coined the term "Japan Fire Blitz," shortening it to Fire Blitz. In boldness of concept and skill of execution, the Fire Blitz resembled Europe's "Big Week" in February 1944, during a remarkably clear weather period that allowed bombing missions nearly every day.

When the notices went up on the bulletin boards on 3 March for Mission #40[†]—that XXI Bomber Command would not schedule any more missions until the 9th—the rumors started. On the afternoon of 9

---

[*] Also rendered as Shitaya.
[†] Also called MEETINGHOUSE II. Most sources erroneously call Mission #40 *Operation* MEETINGHOUSE.

March,[*] the flyers were itching to find out if the rumors were true, with loud bravado replaced by hushed anticipation.[79]

As LeMay and his staffs briefed the B-29 crews, many decided that the Old Man had either gone nuts or was pulling their legs. Some of the more experienced and thoughtful airmen stroked their chins and muttered, *"somebody finally figured it out."* At different times throughout that long, tropical afternoon, LeMay and his briefing officers laid out the details of the mission. They confirmed the worst of the rumors—low-level night firebombing. The fliers were told that they were going to bomb *as individual aircraft* from between 5,000 and 8,000 feet. Altogether, XXI Bomber Command would carry over 2,000 tons of incendiaries to Tokyo.

Silence turned to consternation when they learned they were to leave behind their machine guns, their ammunition and their gunners, except for the tail guns. The pilots then persuaded their briefers to keep the *gunners* because pilots could not see the inboard engines from where they sat. Without the gunners, the flight deck couldn't tell if an engine was smoking because of an engine oil leak or because the wing was on fire.

Resigned to their fates, the navigators, bombardiers, flight engineers, and radar operators attended their specialized briefings as usual. All were certain that this was to be their last mission. Despite a taste of finality, all the crews expected to fly the mission.[80]

When Jack Catton of the 73rd Wing entered the briefing room, his bombardier, Robert Canfield, said, "wait till you see this. It will kill you!" Catton long remembered his amazed reaction to the bombing altitude—7,000 feet.[81]

That afternoon, ground crews loaded nearly fifty thousand incendiary bombs into many[†] Superfortresses (an average of 6.6 tons in each—nearly twice the bomb load carried on any earlier mission). That evening *some* gunners sat in their stations disarmed; others sat at their stations fully armed. Many aircraft commanders and squadrons honored the "no guns or ammo" order—but not all.[‡] By 5:00 PM most of the crews of the mission's 334 Superforts were at their planes. They performed their usual preflight checklist with its multitude of tasks

---

[*] Senior officers had been briefed as early as 6 March.
[†] Guam-based aircraft needed a bomb bay fuel tank to reach Japan and return.
[‡] At least half of the B-29s on the first Tokyo fire raid carried guns and ammunition *against* express orders.

essential to staying alive during a 2,000-plus-mile round-trip mission, even though many were certain that this would be their last.[82]

By then, there was to be no going back on the decisions made. Resigned to their fates, the crews—called the Million Kids by the airplane factory hands—got on with it: the navigators received their bearings and checkpoints, the pilots their engine start and runway times, the bombardiers their bomb loads and release settings…and waited.

Still, as LeMay is said to have confided to an aide, "if we lose, we'll be tried as war criminals." Norstad, again visiting from Washington, fired off a message to his public relations staff, telling them to expect a big story…*if* it worked.[83]

Between November 1944 and March 1945, ten attacks on Japan from China and the Marianas caused fewer than 1,300 Japanese deaths. At the beginning of the campaign from the Marianas, planners told XXI Bomber Command they had to destroy just nine targets to give the Americans absolute control of the skies over Japan. In over two thousand sorties, they had not destroyed a single one.[84]

That, everyone hoped, would soon change.

The Fire Blitz Begins

We must not get soft—war must be destructive and,
to a certain extent, inhuman and ruthless.

HH Arnold

The 314th Bomb Wing under Robert Powers launched the first of its 54 aircraft from North Field on Guam at 5:36 PM on 9 March, nearly seven hours before the first air raid sirens would sound in Tokyo and mere hours after the last of their planes arrived on Guam. At 6:15, John Davies's 110 planes of the 313th Bomb Wing rolled down the runway from Tinian. At the same time, O'Donnell's 73rd Bomb Wing from Isley Field on Saipan launched its 161 aircraft. By 8:10 PM, all the airplanes that XXI Bomber Command could put into the air were aloft. LeMay, much to his regret, had to watch from the ground as the planes rose into the gathering dusk, bound for Japan. Normally, LeMay would have flown an important mission like this. However, they had briefed him on the atomic bomb,[*] and that prohibited him from flying over enemy territory.[85]

---

[*] One source has his first briefing taking place in June 1945, when the 509th Composite Group reached Tinian. However, it seems more likely LeMay knew *of* the weapon before then, and got specifics later.

Nineteen-year-old Ono Kimie slept peacefully in his home in the Hamacho area of the Nihonbashi ward, near Tokyo's Main Railroad Station, the famous Ginza, and within sight of the grounds of the Imperial Palace. His home, with eight occupants, was in Tokyo's *shitamachi* (literally, "under-city," the poorest part of town) on the west bank of the Sumida River, just north of Tokyo Bay. Ono's ward—one of the 35 that made up Tokyo—had a population density of somewhere between 80,000 and 135,000 per square mile. Ono slept in what American planners called Target Area 4 of Incendiary Zone I.[86]

Flying seven thousand feet over the blue-black ocean at nearly 290 miles an hour, Walter Sherrell, aircraft commander/pilot of *Southern Belle*, remembered feeling alone on his way to Tokyo. He did not see another B-29 until he was over the city hours after takeoff—a dramatic change from the near traffic-jam conditions of traditional bombing formations. With the battle for the islands still going on, the bombers had to divert seventy-five miles around Iwo Jima and Chichi Jima; that far out and even farther, trigger-happy Navy ships fired on them. On their way to the target, both Radio Tokyo and Radio Saipan serenaded the crews with "Smoke Gets in Your Eyes," "My Old Flame," and "I Don't Want to Set the World on Fire," among other selections.[87]

For some, the music didn't help, for this was an unconventional, dangerous, possibly suicidal mission ordered by a general many of the B-29 officers simply didn't like, and few knew well enough to trust, even if, oddly, the men *did* trust him. Most worried about the flak; others worried about the fighters they would encounter at such low altitudes. Some men prayed that the element of surprise would somehow overcome all those dangers, but the unaccustomed isolation—the loneliness of flying for six or seven hours in the dark without seeing another airplane—was nerve-wracking. One pilot, Charles Phillips, later wrote:

> Hundreds of B-29s were all in the same general area, all headed in the same direction at approximately the same speed. None of us knew where the others were.[88]

Tokyo civil authorities had rounded up and killed all the dogs earlier in March, publicly because of a lack of rabies vaccination serum, but just as likely because of a lack of food for people, let alone dogs. Before this, they had killed all the dangerous and large animals in the Ueno zoo over fears that a bombing raid might spark a breakout. It is *possible* that the meat from all these animals might have ended up in the shops, or on the tables of those undertaking the killings. Food was a desperately needed commodity in Japan by then, and taboos fell away as hunger took hold.[89]

South of the Chiba Peninsula, the armada of Superforts broke into two streams, with the 314th Wing skirting the coast of Tokyo Bay and the 73rd and 313th Wings flying up the west coast of the Chiba. Scattered between altitudes of four and ten thousand feet, the Superforts edged ever closer to their unsuspecting target.[90]

The B-29s caught the early warning network unprepared for a low-level night attack. Dutiful IJN picket boats, just over the horizon from the coast, fired flares as the Superfortresses swept overhead—although *after* at least one bomber wing had crossed overhead. Radio intercepting teams, listening for American chatter, finally heard some just before the flares launched. Then panic struck the headquarters of the Eastern Air Defense Command, because no alarms reached Tokyo until after the B-29s made landfall and the IJN's ships off the coast opened fire. Even as fast as the telephone could ring, the ponderous, over-compartmented Japanese alert system was slow to warn anyone about the approaching bombers.[91]

At 10:30 PM, the urgent, mournful wailing of the *Kushu Keiho*—the early warning air raid signal—shattered Tokyo's relative quiet. About this time, George Simeral in *Snatch Blatch*, flying with Powers, was 100 miles south of the southern tip of the Chiba peninsula, his first landfall on the way to Tokyo. As the Superforts made landfall one by one, the pilots cracked the throttles wide open.[92]

Simeral's target was Ono's ward.

At seven minutes after midnight, the lead planes from the 29th Group of the 314th Wing dropped their bombs on Target Area 4 as they roared overhead at 5,000 feet—182 100-pound M47 firebombs from each aircraft, released at fifty-foot intervals. The bombs burst as they hit the ground—spattering phosphorous-laced jellied gasoline that blazed instantly on the wooden homes and shops—marking a huge "X" on the city of Tokyo for the rest of the bombers to aim on, encompassing an area between the Sumida and Ara rivers containing approximately 400,000 people, an area of about three by four miles. The pathfinders immolated one major target, the Tokyo Electric Power Company, in the first minutes. At that moment, surface winds were between 17 and 25 miles an hour, fanning the flames and drying out the wooden structures, turning Tokyo into a tinderbox.[93]

Bombing planners thought of urban areas in terms of "built-upness," or the ratio of roof space to total area. They measured bomb damage in terms of roof area destroyed. Mission #40's target area had about 40% roof space, compared to a typical American city of about 10%, or European of about 15%. By setting enough *very* hot fires that would spread quickly, the planners believed that several important factories on the periphery of the target area could also be affected, along with the many small shops inside. This was why the pathfinder's release interval was so small. But that tight interval would also start "appliance fires" that required the deployment of fire equipment to fight effectively. In this, the planners were correct, for the Tokyo fire department deployed what it had very early.[94]

Just moments before the pathfinders struck, the howling of the sirens jolted Ono and his family out of bed. The planes he soon saw were flying so low he thought *they* had awakened him. When he looked outside, it seemed as bright as noon, with incendiaries falling all around his house, the street filled with fleeing refugees. Grabbing his sister and a hastily filled rucksack he went out into the street and followed the tramline to the adjoining Ryogokubashi ward.[95]

Over Ono's head, the bombs-away message set the pattern for future reports:

> Bombing the target visually. Large fires observed. Flak moderate. Fighter opposition nil.[96]

As the crowd of refugees ran, they seemed to move faster, growing by hundreds every moment. The flames, fanned by the high surface winds, seemed to gobble up the surrounding buildings. A mother and child running ahead of Ono simply burst into flame. No one stopped to help—to stop was to die. Funato Kazuyo, a 12-year-old girl who only days before had returned home from a country village, years later recalled a continual roaring of flames, bombs and B-29 engines that assaulted her ears.[97]

The waves of B-29s behind the pathfinders carried E-46 bomb canisters with six-pound M69 napalm bombs, nicknamed "Tokyo calling cards," into the target area—1,520 bombs in every plane—and dropped them at 100-foot intervals that started fires hot enough to melt steel.*

---

* There are disagreements in the sources about the bomb release intervals for Mission #40, especially when subsequent attacks didn't duplicate Mission #40s success.

Along with the napalm came small numbers of HE bombs[*] to disrupt the fire mains. Within minutes, hundreds of separate fires set whole wards ablaze before they joined in cataclysmic infernos that grew on their own, yet were fueled by more bombs every moment…an average of nearly ten tons of bombs every minute.[98]

Kimura Yoshitaka, then a 7-year-old schoolboy whose family owned a toy store in downtown Tokyo's Asakusa ward, later recalled:

It was as if we could reach out and touch the planes; they looked so big…The bombs were raining down on us. Red, and black, that's what I remember most.[99]

The flames blew Kimura into the entrance of a big department store while he ran toward the Sumida River and thus escaped the horror of the river, where tens of thousands burned, crushed, drowned or were suffocated.[100]

Each Superfortress wing hit its target from a different altitude and at a different time. The 73rd Wing was the first and the highest at 7-8,000 feet; the 314th Wing was second and the lowest at 5-6,000; bringing up the rear was the 313th Wing at 6-7,000 feet.[101]

A half hour after the first bombs fell, there was nothing anyone could do to save Tokyo. The growing, living *things* of flame engulfing Tokyo—unlike the stationary *cyclonic* firestorm that destroyed parts of Hamburg as it sucked in fuel—were *sweep conflagrations*. Even the most experienced firefighters with an unlimited amount of the best equipment will flee such infernos, the likes of which can overwhelm forest firefighters in seconds when the wind shifts the fire-beasts around.[102]

The brave Tokyo firefighters attacked the first fires, but it was no use. Policemen directed panicked people towards burned-out areas as firemen sprayed the crowds as long as they had water. Both police and firefighters gave up entirely an hour after the first bombs fell as refugees, rubble and flames blocked the streets.[†] And, by then, the fire mains had all failed. The heroic first-responders lost nearly 250 pieces of equipment and 85 firefighters, as well as 500 police in the hopeless struggle against the raging infernos. Kase Isamu, a firefighter on duty at a train parts

---

[*] So few—less than 2% of the bombs carried—that most accounts ignore them.
[†] The Tokyo fire chief later said they had *officially* given up after thirty minutes. Under those conditions, many dedicated firemen just didn't stop until it was *completely* hopeless.

factory, jumped onto a pump truck when the attack began, knowing the job was impossible:

> It was a hellish frenzy, absolutely horrible. People were just jumping into the canals to escape the inferno…

Kase said he survived because he did *not* jump in the water, but the burns he suffered doing his duty were so severe he was in and out of the hospital for the next 15 years. Split-second choices like that determined who lived and who died.[103]

The B-29s reported flak of varying degrees of intensity and accuracy, saying the Japanese light AA gun fire was too low and the heavy AA too high. Still, flak damaged 47 aircraft and *may* have downed an unclear number.[104]

The fires and resulting chaos forced people to rely on their own meager resources. Tsuchikura Hidezo, a factory worker, saved his family and himself by climbing into a water tank on a school roof. Tsuchikura said later:

> The whole spectacle with its blinding lights and thundering noise reminded me of the paintings of purgatory—a real inferno out of the depths of Hell itself.[105]

Simeral and Powers in *Snatch Blatch* climbed to 20,000 feet after the pathfinders marked their target. They could see fires burning on the west side of the Sumida River that subdivided the target area. As they watched, Simeral recalled seeing what looked like a Fourth of July sparkler—a trail of white light (the incendiary streamers after bursting from their canisters), followed by a splash of orange (the bombs exploding). As the M69s descended, the burning cotton streamers used to slow their descent were visible from as far away as Osaka.[106]

The 53rd *Sentai* dutifully rose and attacked the B-29s, and claimed 12 destroyed and 11 damaged B-29s to a loss of three fighters. Under the conditions of the raid, they hardly needed the radar sets that they had discarded. Most sources deny the 53rd downed *any* B-29s that night, though one source credits it with two. Fighter defense remained weak throughout the three-hour raid, with the Americans reporting seventy-six sightings and forty attacks, usually when Superforts were in searchlight beams. Superfortress crewmen thought the interceptors worked without benefit of radar, being guided in solely by searchlights.[*107]

Ohtaki Masaharu, 13, fled his family's noodle shop with a friend. Firefighters stopped them from going further, so they headed towards Tokyo Bay, but other first responders ordered them back. The boys crouched in a factory yard, waiting as flames consumed their neighborhood:

> We saw a fire truck heaped with a mountain of bones. It was hard to understand how so many bodies could be piled up like that.[108]

Separate fires joined into one roaring blaze in many areas, becoming several howling 1,800-degree furnaces, sweeping the city, fed by a steady stream of bombers unloading even more canisters of fire—nearly four million pounds of incendiaries.[†109]

A French journalist, Robert Guillain, reported on the scene:

> Bright flashes illuminate the sky's shadows, Christmas trees blossoming with flame in the depths of the night, then hurtling downward in zigzagging bouquets of flame, whistling as they fall. Barely 15 minutes after the beginning of the attack, the fire whipped up by the wind starts to rake through the depths of the wooden city.[110]

Ono and his sister resolved to jump into the Sumida River to escape the flames. But as they reached it, they could see the dead and dying already choked the waters. They went back the way they came, hoping to find some water and protection.[111]

Sherrell in *Southern Belle*, some three hundred miles behind the pathfinders, noticed a glow to the north. His navigator said they were still an hour away from the target—the red glow. As he flew over the inferno, he thought he could smell the pungent odor of burning pine. Other crewmen would later recall the reek of burning flesh.[112]

Rejecting the already packed steel-and-concrete Meiji Theater, Ono and his sister happened on a basement. As they huddled in the scant

---

[*] An observation verified by postwar investigation.

[†] In contrast, the largest German raids on British cities dropped no more than a hundred thousand pounds of incendiaries; the British dropped a maximum of two million pounds in a single raid.

shelter, it filled with water from broken water mains, and more refugees entered. The horror was nothing like Ono had ever seen, with people's bodies covered in burns, their eyes swollen shut, their hair burned away, as they suffered, suffocated, retched, and many died.[113]

Unchecked, the blazes outstripped their marked boundaries, and stopped only when they could no longer feed. The Dantean inferno raged on as many smaller fires merged into larger ones, creating a scattered 15.8-square mile crematorium.* As fire reached Ono's basement sanctuary, the roof of the crowded Meiji Theater caved in, crushing many unfortunates inside. Those crushed were lucky—the rest suffocated as fire sucked the air out of the ruined building. The flames of the rampaging fire-dragons roaring around their basement sanctuary nearly suffocated Ono and his fellow refugees. Not far away at the telephone exchange in Sumida, operators who stayed at their posts throughout the attack saved themselves by throwing water at the window frames—the only wood on the outside of the structure.[114]

Looking down, several Americans described it "was like looking into the mouth of Hell." Some flyers had been on massive B-17 and B-24 raids in Europe, and had never seen such destruction spread over such a large area in such a short time. Savage air currents created by updrafts launched one B-29 2,000 feet up in two seconds. Gordon B. Robertson from the 29th Group fought to keep his plane level at 5,000 feet while releasing the bombs. However, savage air currents created by drafts hurled his plane up, and he tossed around "like a cork in water during a hurricane" before his aircraft flipped upside down. He regained control after a long power dive. Even the circling controller planes felt the heat at 10,000 feet; at 6,000 feet some crews had to wear their oxygen masks. Some crewmen claimed the skin of their airplanes grew too hot to touch.[115]

For nearly three hours, the Superforts dumped their firebombs into the howling purgatory. When they dropped their payloads on and around the inferno, they didn't need to aim. Planes hours behind the leaders just headed for the brilliant glow on the horizon. Eighty-one percent of the 279 B-29s that hit Tokyo† were over the target between midnight and 1:30 AM; the rest hit at about 2:00. The last B-29s departed shortly after 3:00 that morning.[116]

---

* Some sources state that the fire zone became one big fire, but post-strike photos prove otherwise.
† Twenty aircraft of the 504th Group/314th Wing hit secondary targets; 29 aircraft aborted.

Tsuchikura Hidezo and his two children took refuge on the roof of the Futaba School. For an hour and a half, they doused each other with water as their clothes caught fire. The flames receded from the area, leaving only Tsuchikura, his children, and twelve others alive on the roof. He later wrote:

> ...The entire building had become a huge oven three stories high. Every human being inside the school was literally baked or boiled alive in the heat. Dead bodies were everywhere in grizzly heaps. None of them appeared to be badly charred. They looked like mannequins, some of them with pinkish complexion...But the swimming pool was the most horrible sight of all. It was hideous. More than 1,000 people had jammed into the pool. The pool had been filled to its brim when they first arrived. Now, there wasn't a drop of water, only the bodies of the adults and children who had died.[117]

Not even the *Gobunko*, a bunker complex built into the hillside near the Fukiage gardens within the Imperial Palace compound for protecting the royal family, escaped harm. As the raid was just beginning, the Showa Emperor Hirohito was in his underground command post, half-asleep and awaiting two important phone calls—one about the IJA's final takeover of the puppet government in Saigon, Vietnam, the other about the birth of his first grandchild.

Unable to sleep because of the sounds of the raid, the Showa remained safely sequestered inside his bunker. Eventually, the inferno's high winds reached the palace grounds, dropping burning embers on the shrubs and camouflage surrounding the bunker, starting several fires. Palace guards and other staff had only pails of water and branches to put out the flames, and, as they struggled to do so, its acrid smell infiltrated the space where Showa's family had taken refuge.[118]

Finally, by about 8:00 on the morning of 10 March, the largest fires burned themselves out...there was no more fuel. A pall of smoke and gas over three miles high rose into the chilly dawn. It was days before substantial relief reached the center of the fire area. Days passed before makeshift hospitals and aid stations could treat casualties. Funato Kazuyo's family searched in vain for medical attention for three of their five children all the next day, finally abandoning the city. The only military rescue unit sent to Tokyo comprised nine doctors and 11 nurses.[119]

"Stacked up corpses were being hauled away on trucks," Sasaki Fusako later recalled,

> Everywhere there was the stench of the dead and of smoke. I saw the places on the pavement where people had been roasted to death. At last I comprehended first-hand what an air raid meant.[120]

Kiyooka Michiko, a 21-year-old government worker living in the Asakusa ward, survived by hiding under a bridge:

> When I crawled out[,] I was so cold, so I was warming myself near one of the piles that was still smoldering. I could see an arm. I could see nostrils. But I was numb to that by then…The smell is one that will never leave me.[121]

Survivors remember the hush as dawn broke over a wasteland of corpses and debris, studded by bathhouse chimneys and the posts and frames of what remained of drill presses and milling machines of what had been small factories. Ishikawa Koyo, a police photographer, captured the carnage of charred bodies piled like blackened mannequins, tiny ones lying beside them. One survivor recalled, "it was as if the world had ended." Her father sheltered her under his body. Those on top of him burned or suffocated. All her family survived.[122]

A fire captain, describing the scene at the Ryogoku Bridge, wrote:

> [A] forest of corpses packed so closely that they must have been touching as they died. They had returned to humanity's carbon existence, crumbling at the touch…[123]

With an average population density of 103,000 persons per square mile, the death toll* from Mission #40 was somewhere between 80,000 and 150,000—more than either of the atomic bombings. The injured reached a quarter million, probably more. The raid destroyed over a quarter million buildings—about one-fourth of Tokyo's total—leaving a million and more people homeless. There was little rubble left in the burned-out area; only an occasional fire-resistant building, scarred by the heat and burned to a shell, remained. The fires hopelessly ruptured the

---

* A floral monument in downtown Tokyo honors 105,400 confirmed dead. The count can never be accurate because the tidal rivers bounding the area carried thousands out to sea before they could be accounted for, and many were reduced entirely to ash.

water mains, clogged narrower streets with rubble, and left the air choking with ash. The fires knocked out electricity to the central wards, beyond Japan's means to repair it. Less than three hours were enough to ruin a city of seven million, which is approximately the population of modern-day New York City. Takeuchi Tokuji of the Ministry of Interior said: "It was the great incendiary attack on 10 March 1945 on Tokyo which definitely made me realize the defeat." The firefighters of Tokyo stopped drills, their morale broken.[124]

Tokyo, Morning, 10 March 1945

Miraculously, the Earthquake Memorial Hall that housed thousands of jars of remains from the 1923 disaster, was untouched by the flames. Most of Tokyo's crematoria were destroyed; burial parties resorted to open-air mass cremations.[125]

Ignited by the 142-minute air attack, the fires wiped out five of Tokyo's 35 wards: Fukagawa, Honjo, Asakusa, Joto, and Nihonbashi. Fukagawa counted about 30,000 dead between them, Honjo another 25,000, Joto over 13,000. The strike torched half of six other wards; eighteen more wards suffered more moderate damage, while few escaped completely. The raid incinerated 40 percent of the capital, destroyed nearly 100 fire stations and 65,000 feet of hose. "Even to her own people," recalled reporter Kato Masuo, "Tokyo's once beautiful face had become unrecognizable and misshapen." LeMay wrote it was "the most devastating raid in the history of aerial warfare."*[126]

American casualties were as LeMay had predicted: about 5%. Of the 299 B-29s that reached the target, they lost fourteen aircraft. At least three were lost to antiaircraft fire, and one, perhaps, to aerial ramming. The rest were to the violent updrafts of the fires or maybe the guns of the 53rd *Sentai*; two were to weather, three ditched in the sea, and one crash-landed on Iwo Jima. Ninety Americans were lost in Mission #40.[127]

The American media went wild. The Milwaukee (WI) *Sentinel* for 10 March 1945 blared "300 B-29s STRIKE! TOKYO AFLAME!" above even the title banner. Below it was the extended story of the dramatic capture of the Ludendorff Bridge at Remagen, which happened on 7 March. In Minnesota, the Twin Falls *Telegram* announced "300 Superforts Fire 10 Square Miles of Tokyo," putting their headline underneath a story announcing the juncture of the US First and Third Armies in Germany because of the Remagen bridge. Montana's Butte *Standard* headline read: "B-29s Turn Tokyo Into an Inferno," again subordinating the firebombing to the news out of Germany.[128]

The Japanese press, meanwhile, both minimized and amplified the effects of the Tokyo attack. While admitting that a "larger number" of B-29s had reached Tokyo, their figures were about 130, and claimed they shot fifteen down against "negligible" damage. Radio Tokyo that day claimed over 100 B-29s shot down in a "glorious victory." Referring to the raid as a "blind," "terror," and "slaughter bombing" attack, the government never allowed the Japanese media to report how devastating and deadly any bombing raid was, all the while decrying the horrors they delivered. Some compared LeMay to the Emperor Nero.[129]

Post-strike photos showed the fires destroyed 18 percent of Tokyo's industrial area, 63 percent of the commercial area, and the hearts of the most congested residential areas. The XXI Bomber Command's intelligence officers struck off twenty-two industrial targets. Flames also destroyed 449 out of 857 first aid stations, 132 of 275 hospitals, and 97 of 196 maternity clinics. Half of Tokyo's burn patients died within hours of arriving at surviving facilities. Dressings had to be used again and again, often without proper sterilization. The resulting infections killed thousands more.[130]

---

* At this writing, it still is.

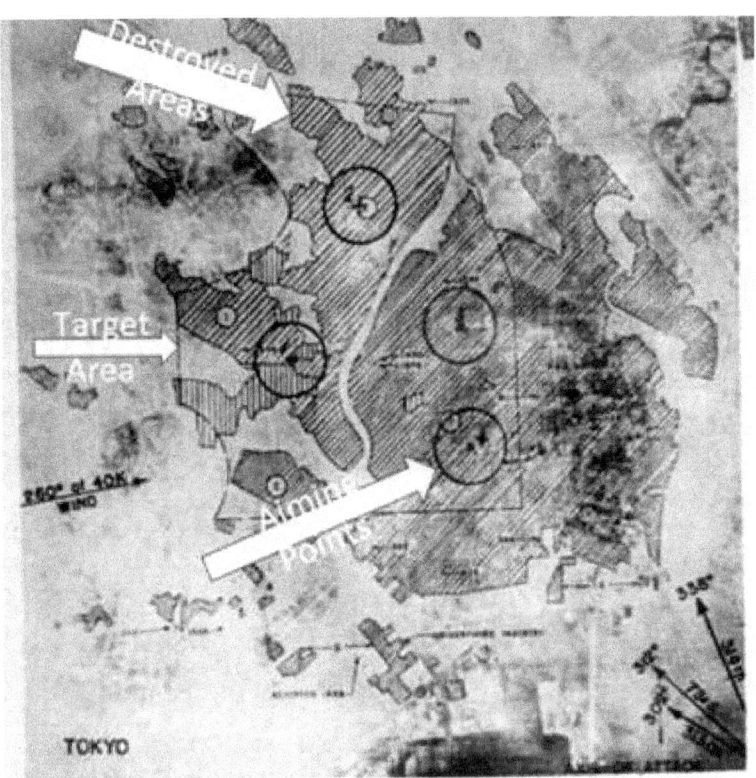

**Post-Strike Composite Photo of Tokyo Raid Effects
from Lemay, Phase Analysis Report**

The Japanese government reaction *out* of the public eye, however, was part stupefaction and part terror. Evacuation of children sped up, extending to 5- and 6-year-olds and their teachers. Authorities practically abandoned civil defense training as thousands of terrified citizens fled to the safety of the countryside. Ugaki, in his diary, called it "a considerable fire" started by 130 B-29s, but he wrote little more about it as the fighting on Iwo Jima preoccupied him.[131]

An official of the Home Affairs Ministry later reported, "We were instructed to report on actual conditions. Most of us were unable to do this because of horrifying conditions beyond imagination."[132]

This was only the beginning....

The Fire Blitz Goes On[*]

Tokyo had dropped through the floor of the world and into the mouth of Hell.[133]

Curtis E. LeMay

A s promised, LeMay and XXI Bomber Command kept up a brutal pace of very long missions, both to exploit favorable weather *and* to maintain the sense of shock and awe. Mission lengths were grueling: round-trip flights up to 14 hours long, preceded by four hours of preparation, plus *maybe* another four to debrief the crew, write the reports and service personal equipment...and sleep, whenever and *wherever* the men could get it.

LeMay gave the flight crews a few hours' rest while the maintenance men reloaded and refueled the planes. Some crews took off less than 24 hours after they landed from the Tokyo mission. Some aircrews flew four missions in as many days—twice the pace as before Mission #40.[134]

---

[*] A great deal more has been written about Mission #40 than the subsequent fire missions. This hopefully explains the disparity here.

Nagoya

On 12 March, only 29 hours after the last Superfortress returned from Tokyo, LeMay launched 310 B-29s to Nagoya[*]—Japan's third largest city and its largest aircraft and aircraft engine producer—to do what they had done to Tokyo. But the bomb mix this time was different, as only enough M69 incendiary clusters were available for the pathfinders. By supplying the Superforts' tail guns with 0.50 ammunition as well as 20 mm, LeMay hoped to knock out Japanese searchlights.[135]

Nagoya had mushroomed as a producer of war material. The target area was a triangular wedge of the city with a population of about 70,000 to the square mile in a nearly 40% built-up area. Within the city were the Mitsubishi engine and assembly works, their affiliated units, the Aichi aircraft plants, and two aircraft engine plants under construction. While Nagoya's harbor was small, it was a major transportation center on the main Tokaido railroad and boasted an elaborate system of military highways. Because Nagoya's closely packed industrial and urban areas were extremely vulnerable to incendiary attacks, the defenders peppered the city with heavy AA and machine gun emplacements.[136]

Launching once again as darkness fell, one B-29 ditched soon after takeoff and nineteen others aborted. Two hundred and eighty-five bombers got over the target, dropping 1,710 tons of E-46 clusters for about two hours. This time the 313th Wing hit the target from 7-8,000 feet, followed by the 314th Wing right behind them at 6-7,000 feet, with the 73rd Wing bringing up the rear and bombing at 6-7,000 feet. As they launched, some say LeMay remarked, "the only thing the Japanese have to look forward to is the total destruction of their industries."

The USAAF burned out two square miles of Nagoya, and they *thought* they knocked out eighteen industrial targets, but the attack only slightly reduced the Aichi plant's output. There were several reasons for the *apparent* lack of success at Nagoya. Nagoya was less susceptible to large fires than Tokyo because most of the buildings were newer, using more masonry and other non-wood construction. It also had more and better designed firebreaks and a better organized fire department. Additionally, the surface wind for Mission #41 was negligible, such that the many fires never merged into the sweep conflagrations that engulfed Tokyo.

---

[*] Also, MICROSCOPE II or Mission #41.

The debriefings from Mission #40 created the impression that dropping bombs too close together resulted in wastage. This false impression caused changes in the bomber's intervalometers to release the bombs from fifty feet to a hundred feet over Nagoya, which caused the bombs to scatter more than they had at Tokyo. For the Nagoya strike, the resulting density pattern was too thin to start a Tokyo-like holocaust.

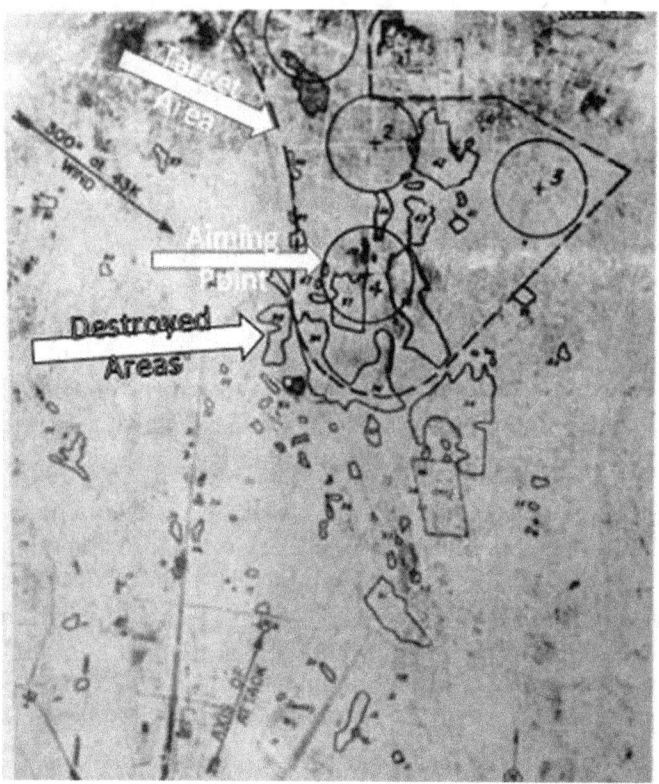

Post-Strike Composite Photo of Nagoya I Raid Effects
From LeMay, *Phase Analysis*

And, the 313th and 314th Wings attacked Nagoya one hour before the 73rd, which reduced the surface heat that contributed to the Tokyo blazes. The planners spaced the Mission #41 aiming points to avoid blacking out the target for late arrivals. They briefed the bombardiers to place their bombs visually near the aiming points to cover the entire area. Only a few planes made a controlled run over the target, and the attempt to scatter the bombs by snap judgment resulted in too wide a dispersal.[137]

Only one B-29 went down during Mission #41, and that was the ditched plane at takeoff.

Flak, despite being described as "heavy…generally inaccurate," damaged eighteen planes, with a reported 97 fighters[*] hitting two more in 43 attacks. A US Navy submarine 150 miles away reported that wood smoke had reduced visibility to less than a mile.[138]

While little information is available on Japanese casualties caused by Mission #41, there were nearly four hundred fires started by the bombers, some of which didn't stop until they hit a fire break. Nagoya might have escaped the broader devastation that Tokyo experienced the night before, but even the smaller scale devastation of that many fires would have resulted in hundreds, if not thousands, of casualties while overloading the hospitals and destroying thousands of homes and other structures. At this point in the war, *any* damage at all was difficult for Japan to repair.[139]

Osaka

Before evaluating the results of the Nagoya mission, LeMay scheduled Mission #42—Osaka[†]—for 13 March. While Kobe was the big deep-water port for central Japan, Osaka was an important coastal shipping center: the small vessels critical to Japan's survival used its extensive network of canals that were lined with warehouses and boatyards.[140]

Japan's second largest city of some three million souls that had yet to be hit by a major air attack, Osaka produced about one-tenth of Japan's wartime total of ships, one-seventh of her electrical equipment, and one-third of her machinery and machine tools. It was a focal point for land and sea transport and a diversified concentration of industry, ranging from steel mills and shipyards to precision tools and machine parts. The Osaka arsenal furnished 20% of the IJA's ordnance requirements. They assembled no airplanes at Osaka, but nearly a fourth of its half-million workers made parts and components for aircraft and engines. Osaka was also a great commercial city and an important administrative center. Because of conscription and the mushrooming of war-industry plants in the suburbs, the city's numbers had shrunk from 3.2 million in 1940 to about 2.1 million in 1945. The shift had reduced the population density to 81,000 per square mile for the central commercial section and adjacent residential-industrial districts without affecting the built-upness—there was still plenty of fuel to burn. The scene of many earlier disasters,

---

[*] These numbers seem improbably high. Japanese fighters claimed one B-29 kill.
[†] Also, PEACHBLOW I.

Osaka had cut firebreaks through congested areas to add to the protection given by its many canals, and had built many modern fire-resistive buildings, but its crowded districts of highly inflammable houses still offered an ideal incendiary target.[141]

LeMay's planes carried the same 6-ton bomb load, but the 73rd Wing had 0.50 ammunition for lower forward and aft turrets and for the tail guns. Thanks to heroic efforts by maintenance crews, the command put up 301 B-29's for a late afternoon launch, but there were concerns among senior commanders:

> I am pushing my maintenance entirely too far. While I realize the importance of striking while the iron is hot burning yourself in the process does not make sense. I have tired crews, tired mechanics and tired airplanes. Am checking and re-checking to be sure the airplanes which do go out are in top shape.[142]

Reports of the Tokyo and Nagoya missions raised doubts about the tactics used at Nagoya, and operational planners tried to reproduce the pattern which had worked so well in Tokyo. They prescribed no specific method to achieve a higher concentration in the target area, although they briefed crews to check their position carefully before releasing the bombs. The planners returned the intervalometer settings to those used for Tokyo.[*]

When the force of 274 planes that got over Osaka found an 8/10 cloud cover, it had to resort to radar bombing. This proved an advantage rather than a handicap. Unable to sow their bombs by sighting visually on pathfinder fires, radar bombing forced bombardiers to drop after a controlled run, releasing on an offset aiming point. With this technique, the B-29s achieved a thicker and more uniform pattern than had been possible with the impressionistic methods used at Nagoya.[143]

The incendiaries did their job, charring about 60% of the theoretical fire zone. Forty-eight pieces of fire equipment burned where they sat, immobilized by a lack of gasoline. Columns of smoke reached to over 8,000 feet, so dense that bomber crews flying through them couldn't see the outboard engines. Fifteen B-29s aborted, possibly because of the pace of operations.[144]

The powerful concussive force caused by the Osaka raid batted some B-29s out of the air. One shock wave blew the ironically named

---

[*] Once again, sources are contradictory; either fifty or a hundred. It is possible that the pathfinders were set to fifty and the rest to 100, but this is unknowable.

Topsy-Turvy of the 468th Bomb Group 5,000 feet straight up. Flipped upside down, the plane fell 10,000 feet before the pilot got it under control.[145]

Post strike analysis showed conclusively that the Tokyo raid had not been a fluke. The 1,732.6 tons dropped on Osaka in about three hours wiped out 8.1 square miles in the city's heart. The chief commercial district was over 75 percent destroyed. Fires kindled in the industrial sections destroyed 119 major factories and damaged some 2,000 small manufacturing plants. Damage to communications, urban transport and storage facilities was extensive.

As the flames spread, first responder morale collapsed. Casualties mounted as people suffocated in makeshift shelters or burned while trying to run through the flames as they had at Tokyo. The records of the Osaka fire department listed nearly 4,000 dead, over 600 missing, over 8,000 injured, and 600,000 left homeless. While the bombing destroyed over 130,000 houses, it *damaged* less than 1,400, a testament to the fury of the fires. While sufficient housing probably remained to take care of essential workers, a holocaust of such dimensions was bound to increase absenteeism and disrupt industry. Over 400,000 fled Osaka.

The cost to XXI Bomber Command was very light. Two B-29's were lost—one at take-off—and thirteen damaged. The early warning network alerted Nagoya soon enough to assemble an interceptor force, but it made only a feeble effort. The B-29s reported only forty individual attacks at Osaka, and no Japanese fighter scored a hit.[146]

Back-to-back-to-back major raids generated tremendous news coverage in America. "Strategy Shifts to Area Bombing," the New Orleans *Times-Picayune* headline blared. An Associated Press dispatch announced:

> The mounting fierceness and relentlessness of the Superfortress incendiary raids on Japan's industrial centers strongly indicates that the B-29 command has embarked on an all-out campaign of area rather than strategic bombing.

This trend alarmed Washington officials, who sent a cable to LeMay on 14 March, saying "guard against anyone stating this is area bombing." However, Washington offered *no* suggestion what label to use to disguise what it clearly *was*.[147]

Yet, the pace of operations and shortages of some vital parts were critical. As soon as a battered plane touched down, mechanics yanked off damaged bomb bay doors, repaired them, and rehung them, often in as little as twenty hours, because there were so few available. To motivate service crews, LeMay ordered strike photos pinned up on bulletin boards, so everyone could share in the results. The strenuous pace likewise took a toll on the pilots, navigators, and bombardiers, who swallowed both sleeping *and* pep pills as needed. One officer reported:

> A marked loss of appetite was noted due to fatigue caused by irregular eating and irritated nerves...It was hard for us to relax and sleep as most of our rest must be acquired during the hours of daylight when the temperature and the humidity were the greatest.[148]

By 15 March, Tokyo buried over 77,000 remains in parks, school grounds...anywhere holes could be dug. It was twenty-five days before workers removed all the dead from the ruins. Workers packed many cremated victims in the Great Kanto Earthquake shrines, but they interred most in hundreds of unmarked mass graves wherever they found open space.[149]

Now Nagoya and Osaka began burying *their* dead in *their* shrines and open spaces. But the Fire Blitz continued...

Kobe

In February, LeMay calculated his needs based on 735 sorties per wing per month carrying 4-ton loads, with bombs at a ratio of 60% high explosives and 40% incendiaries—conventional requirements. For the three wings available in March, this meant he would need 3,528 tons of incendiaries of all types. Now, in three missions, he had dispatched 948 sorties loaded with an *average* of 3,900 tons of mostly incendiaries per mission, well over a month's supply drawn only from the stocks of M47s and M69s.

The XXI Bomber Command was running out of those optimal incendiaries.

Kobe was Japan's sixth largest city, across the bay from Osaka, about 25 miles southeast. It was a long, narrow industrial city squeezed between Osaka Bay and mountains that drop almost to the shore. Kobe was Japan's most important overseas port, and a focus of inland transportation. To either side of Kobe's harbor and its commercial area lay important heavy industry installations. The Tokaido railroad to Tokyo divided Kobe's urban area lengthwise, with the shipyards, industries, and central business district on the bay side, and the residential section on the other. Crowded near the waterfront were construction and repair facilities for naval and merchant vessels, ordnance and steel mills, and small factories feeding the big shipbuilding complex. The Mitsubishi and Kawasaki marine engine and electrical equipment yards were in Kobe. Many ships came to Kobe to have equipment installed. The city also provided equipment to other shipyards.

Kobe cut firebreaks in 1944, but they were intended to protect individual targets rather than prevent fires from sweeping over sizeable areas. It had been the target of a small test incendiary raid on 4 February, and had caught a few stray HE bombs during a precision attack on nearby Akashi on 19 January, but remained practically a virgin target.

On 16 March, LeMay launched three hundred thirty Superfortresses against Kobe for Mission #43[*]. LeMay's operations officers, convinced that visual distribution methods were unsatisfactory, again changed some tactical details. The field orders called for bombardiers to make a controlled radar run over the target before making visual corrections and to apply such corrections only to their sighting on the aiming point: they were *not* to spread bombs visually. Kobe, with its long irregular waterfront, provided an excellent radar target. The planners had come to appreciate the value of greater concentration to insure the merging of individual fires, changing the flight schedules to reduce the duration of the attacks, and plotting the aiming points in closer patterns.

Because of low stocks of M69 napalm bombs, they had to substitute less-effective M17A1, a 500-pound cluster of 4-pound thermite canisters. With 110 individual bombs per cluster, the M17 would achieve a wide dispersion, and the thermite missiles would be effective in the dock and heavy industry areas. However, they were *not* as destructive as petroleum bombs against flimsy dwellings. The best thing about the M17 was its abundance.

Though morale was soaring in the XXI Bomber Command for the first time in months, the crews were getting tired: there were 21 aborts on the Kobe raid, and not all for mechanical malfunctions. Three missions in four days—interrupted by two Japanese air raids on Saipan—meant that some flyers got four hours of sleep in 36 hours...or less. While the maintainers *could* get perhaps another *hour* of rest, the flight crews could not. Sky-high morale or not, it was telling on the

---

[*] Also, MIDDLEMAN II.

airmen. That, *and*, the crews were genuinely afraid of Kobe, the biggest flak center in Japan; bombing it at less than ten thousand feet sounded like suicide.[150]

Three hundred and seven bombers dropped some 2,328 tons of incendiaries in just two hours and eight minutes that burned about 20%—2.86 square miles, nearly 2,000 acres—of Kobe. The bombing inflicted significant damage to both major shipyards, and destroyed many large waterfront cargo and military storage warehouses, some of which were Japan's last word in modern reinforced concrete construction. Japanese reports showed about 500 industrial buildings destroyed, more than a hundred damaged, and nearly 66,000 houses destroyed. Police reported over 2,600 dead or missing and 11,000 injured, with nearly a quarter million rendered homeless. But it might have been worse: many thermite bombs crashed through flimsy wooden roofs and buried themselves in the earth, diminishing their effect. Even if the fires started in a smaller area and in smaller numbers and were more conventional, Japanese civilians and first responders struggled to survive against the flames.

And the Japanese were responding: bomber crews described antiaircraft fire over Kobe as "medium to heavy, meager to intense, inaccurate." Fighter opposition was increasing: "medium to heavy, unaggressive—93 attacks," with reports of over 300 fighters sighted—an impossible number. KI-44s from the 245th *Sentai* based at Taisho airfield in Osaka met them over the target. Two rammed a B-29 and survived—it was one pilot's second ramming,[*] having successfully brought down another B-29 with a collision three nights earlier—but the Americans counted no bombers lost to fighter opposition.[151]

Some Japanese ground crewmen at an Osaka airfield saw a spot of light streaming like a shooting star into a B-29 that was lit up by two searchlights. Seconds later, a huge fireball erupted. Major Furukawa recalled the destruction:

> The roaring flames from Kobe city shot skywards, inviting boiling clouds and strong winds. With the wind blowing from the west, there was a menacing atmosphere prevailing all over the area.[152]

According to the Americans, Japanese fighters shot down no B-29s that night. But one KI-44 rammed *Z-8* two miles north of Kobe, bringing

---

[*] Ramming attacks were not necessarily suicide missions, and not all B-29s that were rammed were lost.

it down near a POW camp, where the Japanese captured two crewmen. An Osaka court martial convicted both in a two-hour trial and executed both on 18 July.

In the POW camp, Friar Marcian Pellet, a Franciscan missionary interned with the POWs, saw the B-29 falling out of the sky at about 4 that morning. One victim of the crash was the co-pilot, Robert Copeland. "When the Jap soldiers failed to bury the bodies, we internees secretly buried them," Friar Pellet wrote to Copeland's mother after the war:

> We later erected crosses with their names and marked the spots clearly. Rest assured that we fathers in the camp prayed at the graves, and also in our masses, for the men.[153]

The Americans had reason to be afraid of bailing out over Japan. Several Superforts bombed secondary targets just to say they did something; some bombed targets that weren't even tertiary for the same reason. Several flyers quit their flying status.*

Nagoya Again

Mission #44, the raid on Nagoya on 19 March,† was a departure from the earlier formula. Every third plane carried a couple of 500-pound HE bombs to disrupt organized firefighting, and loaded up to capacity with such incendiaries as were available in the fast-dwindling stocks—M69s for the 314th Wing, M47's for the 313th, and M47s and 500-pound M76s for the 73rd.

Of 313 B-29's dispatched, 290 reached Nagoya to drop 1,858 tons of bombs, burning out another 2.95 square miles of Nagoya's city center. Smoke and searchlights bothered bombardiers, but by using radar they could blanket a considerable part of the city. Many important individual targets were damaged, including the Nagoya arsenal, the freight yards, and Aichi's engine works, but the Mitsubishi plants escaped with minor damage. Over 300,000 people fled Nagoya after this attack—close to a third of her prewar population.

But again, there were Superforts that simply didn't make it to the target for a variety of reasons. They often cited the weather, but usually they were just too tired, or too afraid of the law of averages, or just too

---

* Legally possible for all American, British and German flyers.
† Also, Nagoya II or MICROSCOPE III.

fed up with the heat and the stench of burning cities to keep dropping bombs.[154]

On 19 March,* the Showa Emperor Hirohito left the palace grounds for the first time since October to inspect parts of Tokyo, getting out of his car several times to gaze at the empty areas where buildings once stood, and at a bridge where many thousands had sheltered...and perished. There are no accounts from any witnesses, but Ishikawa Koyo captured the image below at the Tomioka Hachimanga Shrine in Fukagawa.

Showa Inspecting Tokyo (WWII Multimedia Database)

It is said at one point, he wept. Some believe that, with that inspection, he realized that the war was hopeless for Japan.[155]

The First Ten Days of the Fire Blitz

For Japanese morale, some say those first ten days were the beginning of the end. Within that period, with an average of 380 aircraft assigned, XXI Bomber Command had flown 1,595 sorties—three-fourths of the total flown in all previous missions—with the 9,365 tons of bombs dropped three times greater than the weight dropped before 9 March. Not only did LeMay increase his loads, but *most* of his bombers, even with poorly trained radar operators, could hit the target with accuracy, effectively *doubling* the power of his force.

---

* Some sources say 18 March.

XXI Bomber Command had turned 32 square miles of Tokyo, Nagoya, Osaka, and Kobe to rubble and ash. The five raids killed over 110,000 people, injured another 60,000 or more, and left over two million homeless, many of whom had no choice but to flee the cities. "I was not happy, but neither was I particularly concerned, about civilian casualties on incendiary raids," LeMay recalled:

> I didn't let it influence any of my decisions because we knew how the Japanese had treated the Americans—both civilian and military—that they'd captured in places like the Philippines.[156]

The cost in aircraft—21—and crewmen had amounted only to 0.9% of those taking part, a loss ratio far below those of daylight missions, even in Europe. And although the strain of five 14-hour-long missions in five days on both flight and ground personnel had been tremendous, neither group showed signs of cracking on a *large* scale.* Maintenance crews suffered from severe physical exhaustion, but most recovered rapidly. Combat crews—including those who had flown all five missions—despite feeling fatigued, finished the ordeal in *generally* good physical condition, but *with* frazzled nerves. After months of small results and heavy losses, airmen who had become discouraged in the dull routine of tallying combat missions waiting for rotation took on a more aggressive spirit. That said, some airmen in the 73rd Wing were on the verge of exhaustion, having been bombing for nearly four months without major respite.[157]

Arnold's $3 billion dollar gamble was paying off.

The tactical implications of the Fire Blitz were clear enough. By bombing individually from low altitudes, B-29 crews had vastly improved the performance of their planes and their bombing. Bomb loads had increased and engine strain had diminished, getting more sorties per plane. Radar, sufficiently accurate for area bombing, had taken some of the curse out of Japan's weather. Neither flak nor fighters had inflicted serious losses at night. Most important of all, Japan's urban areas had proved highly vulnerable to incendiary attack. The AAF couldn't use area bombing efficiently against all targets, but under

---

* As a rule of thumb, each hour in the air required three man-hours of ground maintenance, more if there was battle damage or a malfunction. The grueling frequency of the sorties gave a plane's 3-man ground crew just 32 man-hours to perform 42 man-hours of maintenance, which meant that some of that maintenance just didn't get done.

suitable conditions, they were so highly successful that the doctrines of strategic bombardment underwent a radical change.

Norstad hosted a press conference at the Pentagon on 23 March, declaring:

> It is very doubtful that such a high cost has ever previously been inflicted upon any people in a single eight-day period in the whole history of warfare….The mission of the Twentieth Air Force is the reduction of [the] Japanese ability to produce war goods…by any and every means at the disposal of the attacking force.[158]

Hansell's clean world of high-altitude daylight precision bombardment was disappearing, but he wouldn't let it go. He sent a congratulatory note to LeMay, saying,

> The decision to go into Japan at such low altitude, at night, was certainly a very courageous one, but obviously it was a correct one….the successful bombing and burning of the great population centers has certainly offered a tremendous contribution.

> But…

> Personally, I believe we will have to return to daylight bombing of selected targets, before we beat the Japanese down to the level needed.[159]

Both the Americans and the Japanese carried out a battle of conflicting leaflets throughout the Fire Blitz. LeMay's Superfortresses delivered messages warning the Japanese which cities they would bomb, plus the assurance that the targets were their industries, not the civilians who operated them. These messages explicitly told the readers to flee the cities, which they did when they could.

While Japanese city-dwellers could pick up American leaflets by the bushel basket, they could get a smaller number of Japanese leaflets as well. These urged their city-dwellers to keep working in their factories, promising victory over the "cowardly Yankees." Which messages were believed is a matter of pure speculation, because no English-language oral histories address them. It *is* notable that urban populations fell precipitously in the spring of 1945. But whether that flight was because of the bombings' breakdown of services—amplified by a lack of food—or the messages themselves, is unknowable. LeMay himself didn't think the leaflet campaign had any effect, but did note that the largest number of civilian casualties—which he always regretted—was in the first Tokyo raid before the leaflet campaign started.[160]

The March raids struck a blow to Japanese morale, as the US Strategic Bombing Survey reported:

> …[T]he March series of attacks struck a penetrating blow at Japanese morale. The war had been driven home to the people. A great number of small subcontractors were completely destroyed in addition to the industrial capacity of many large war industries. Hundreds of thousands of people were casualties or were displaced from their homes and work. Serious problems in emergency housing, medical care[,] and food distribution were created for the overtaxed [Japanese] government and municipal agencies. The productivity of many industries, untouched by the bombs, was appreciably reduced because of labor displacement, disruption, and destruction of transportation and other service facilities. Japan suddenly was faced with a weapon which was capable of rapidly disintegrating her over-all home economy[,] and against which she could offer no effective means of defense.[161]

Yet, LeMay pressed on…

Success in war is often a matter of surviving, but never in the course of WWII in the Pacific had there been such an unambiguous air power success as the Fire Blitz. There was an interruption of major fire raids after the Fire Blitz, in part to replenish the supply of incendiaries, but also because the B-29s were being tasked to assist the invasion of Okinawa. That began 27 March 1945 with the bombing of the kamikaze airfields on Kyushu that continued until mid-June. Sea mining operations by the 313th Wing, that the AAF specifically trained them for, also took place while LeMay waited for more firebombs.[162]

Of the Okinawa support missions, which the Japanese seemed to expect, Ugaki wrote:

> The army seems completely defenseless in protecting aircraft on the ground. In view of the above, I further warned my command.[163]

What may be most important was that by the end of the Fire Blitz, LeMay, in his *Phase Analysis* report, had not *completely* abandoned his earlier ideas about formations and high-altitude bombing because area raids were not suitable for all targets. He later wrote that they still conducted high-altitude precision attacks. By 15 April, LeMay's intelligence people estimated that Japanese aircraft production had fallen to 40% of its peak.[164]

And now LeMay, ever the planner, was thinking ahead to what he felt would be the critical month of September, 1945, when two more wings and XX Bomber Command would reach the Marianas. With those additional forces, he calculated he could increase the number of sorties from April's 3,000 to 6,700. *Except*...with casualties, simple attrition, and crew rotation, he'd be over 900 crews short. Arnold told him they could produce no more aircrews any faster. Converting Dolittle's Eighth Air Force to B-29s (to be based on Okinawa) would take as much as a year and suck up not just planes but precious trainers. Even though many of his crews were exceeding their recommended monthly mission hours[*] by over 20% on their own hook, the flight surgeons worried about the strain, additional mistakes their weariness might bring, and the emotional toll. LeMay believed it was worth the gamble to keep up a brisk, if not break-neck, pace.[165]

Neither LeMay nor the USSBS or anyone else understood why the Japanese stayed at their posts...but stay they did. All over Japan, industrial workers sheltered in slit trenches against thousand-plus degree fires, and worked at their factory jobs, making more of whatever their factories made and scavenged for what food they could find. Policemen ate whatever they got from government handouts while their children hid in the mountains. Government workers' precious artifacts, centuries old, went up in smoke while their sons, some of whom they had not heard from in months or years, served the Emperor...*somewhere*. All the while, their Emperor and their leaders told them to have courage, that glorious victories were just around the corner....

Tokyo Again...Thrice

Fire raids continued as the Okinawa support missions wound down, though not with the same high operational tempo as during the first ten days. Lemay launched another fire raid on the Tokyo arsenal complex

---

[*] In addition to a 35-mission rotation standard, B-29 crewmen also had a 60-hour/month limit.

northwest of the Imperial Palace on 13-14 April (Mission #67). Three hundred twenty-seven B-29s dropped 2,139 tons of bombs, burned out 11.4 square miles of that important industrial section, destroying plants that manufactured and stored small arms, machine guns, artillery, bombs, and fire-control systems, and produced smoke columns up to 16,000 feet high amidst thermal turbulence. Witnesses as far as 100 miles away from Tokyo reported seeing the explosions.

On 15-16 April, Mission #68 targeted the Kawasaki industrial area near Tokyo. One hundred and ninety aircraft from the 313th and the just-arrived 315th Wings dropped 1,110 tons of incendiaries and fragmentation bombs, burning out about 4.5 square miles (55%) of the target area. They met clear weather, ideal for Japanese flak crews and increasingly aggressive fighters, to which they lost 12 planes. This raid destroyed an additional six square miles of Tokyo (mostly along the west shore of the bay), 3.6 square miles in Kawasaki, and 1.5 in Yokohama, which was hit by spillage. The two raids destroyed over 200,000 buildings in Tokyo and Yokohama, and 30,000 in Kawasaki. Japanese statistics on Tokyo casualties vary widely, but were much less frightful than those of the raid of 9-10 March. That same night, the 73rd Wing sent 112 B-29s to Tokyo in Mission #69, dropping another 769 tons of firebombs and burning out another 5.2 square miles.[166]

Ugaki's diary entry for 13 April states that there were four hundred B-29s at Guam and Saipan, but it also states that US losses in the two weeks prior were so catastrophic that the Americans *had* to give up soon. He also believed that the kamikazes had killed FDR (who passed on 12 April), "brought down the cabinet," and placed Truman in charge. While the Superfort count was roughly accurate, the rest was fanciful. Even senior Japanese officers, it seemed, still believed in miracles.[167]

There were no fire raids between 16 April and 14 May. This was due, in part, to the diversion of the 314th Wing to aerial mining and counter-kamikaze attacks, and the 315th Wing for their attacks on Japan's petroleum industry, for which that wing was trained and equipped. *And*, once again, LeMay ran out of fire bombs.

But the stream of refugees leaving the ruins of Tokyo and the other cities was continuous. Burial parties declared they were running out of fuel to burn the dead and places to dig graves. Material for markers was also running short. Even clean water was becoming scarce. A Tokyo factory making balloon bombs (*Fu-Go*) to be launched against the United States had electricity only five hours a day, while their workers toiled by candlelight for eighteen.

Nagoya Again...and Again

The 14 May raid (Mission #174) was the first four-wing strike on the northern built-up area of Nagoya, near the 17th Century Nagoya Castle. This industrial-residential district included the No. 1 precision target in Japan, the Mitsubishi Aircraft Engine Works, as well as the Mitsubishi Electric Company, the Chigusa branch of the Nagoya Arsenal, and many lesser war industries. The population density in the nine-square-mile area was up to 75,000 per square mile. Planners scheduled the mission as a 16,000 to 26,500 foot daylight strike, partly to confuse the Japanese defense, partly to see if accurate high-level firebombing was possible.

Of the 529 aircraft launched, 472 dropped 2,515 tons of M69s on the target at altitudes ranging from 12,000 to 20,500 feet. Bombing was supposedly downwind, but smoke blown across the area made it necessary for some planes to resort to radar. The Nagoya fire department efficiently contained some of the 100-plus identifiable blazes in check in the moderately built-up residential sections. Only one large conflagration near the castle got out of hand, but was stopped by a 200-foot firebreak. The many burned-out areas amounted to 3.15 square miles. Mitsubishi's No. 1 engine works lost its Kelmur bearing plant and suffered other damage.

Japanese reaction to the daylight attack was lively. The attackers lost ten B-29s: one to flak, one to fighters, and eight to other causes; and sixty-four damaged. The B-29 crews claimed eighteen enemy planes destroyed, sixteen damaged, and thirty probables during the battle.[168]

LeMay launched another maximum effort—this time 522 bombers—against Nagoya on the night of 16 May (Mission #176), targeting the dock and industrial areas in the southern part of the city, where the Mitsubishi Aircraft Works, the Aichi Aircraft Company's Atsuta plant, the Atsuta branch of Nagoya Arsenal, the Nippon Vehicle Company, and other numbered targets were located. Planned as a low-level attack, pathfinders pinpointed the aiming points with M47 incendiaries and the remaining B-29s dropped four-pound M50 magnesium bomb clusters suitable for the heavy structures in the area.

When compared to Mission #174 on 14 May, Mission #176 illustrates the relative advantages of high-altitude formation day versus low-level night attacks. Fewer planes—457 as against 472—hit the primary target. But the low-altitude, individual aircraft approach allowed a heavier bombload per plane—about 8 tons against 5.3—dropping

3,609 tons in all that started over a hundred fires, burning out 3.82 more square miles of Nagoya and heavily damaged Mitsubishi's No. 5 aircraft works. Because production dispersal efforts were well underway, there was little change in output following the raid, though the Mizuho plant that made engine cowlings turned out only twenty units after it. Because of smoke and thermal drafts, some planes had to bomb from levels much higher than those designated in the field orders, decreasing accuracy. As usual in night attacks, the opposition was weak. Mechanical failures were responsible for the three B-29's lost. Advantages and disadvantages, the losses notwithstanding, about balanced out.

The earlier precision attacks and the fire raids had already burned out twelve of the roughly forty square miles of the built-up urban area, ruining the industrial fabric of the city. The Superforts had destroyed over 113,000 buildings, killed over 3,800 people, and made nearly half a million homeless. Mission #176 finished Nagoya as an area attack target. Nagoya Castle, known as the "castle of the golden dolphins" for its golden *shachihoko*, votive tiger-fish roof devices, and the Hommaru Palace within its walls (then used as a military headquarters), burned to the ground.

Another 170,000 and more people fled the city. Nagoya may no longer have been an incendiary target, but neither was it a fit place to live. At least in the country, one could boil grass and look under logs for grubs to eat…which was more than they could find in the ash heaps of Nagoya.[169]

Tokyo Again…and Again

Planners called the target for Mission #181 Tokyo Urban Area No. 3, a region stretching southward from the Imperial Palace along the west side of Tokyo harbor that included both industrial and residential communities. The palace presented a special operational and tactical problem. Its grounds—just short of a square mile in area and surrounded by a moat—was a convenient checkpoint for navigators, but formed the most effective fire barrier in the Chiyoda ward. That said, the upper political and military echelons ordered LeMay's pilots to avoid dropping bombs on or near it "since the Emperor of Japan is not at present a liability and may later become an asset." In four previous area raids, over 5,000 tons of bombs had destroyed a little over 34 square miles of Tokyo.

The 562 B-29's launched on the night of 23 May (Mission #181)— the largest B-29 attack of the war—included 44 pathfinders; the tactics called for the familiar and previously successful combination of M47 and

M69 incendiaries. There were 520 B-29s over the target[*] for two hours, dropping 3,646 tons of bombs at an average rate of 1,000 pounds a second from altitudes ranging from 7,800 to 15,100 feet. The planned axis of attack was to avoid the heaviest ground defenses, but flak was intense, though fighter opposition was light. Seventeen B-29's were lost (four operational losses) and sixty-nine damaged. The weather was bad with 9/10 cloud cover, and bombardiers had trouble with smoke and the searchlights, but they got enough bombs into the area to burn out another six wards—5.3 square miles south of the Imperial City. Aircraft density was so high there were traffic jams on the return leg.[170]

On 25-26 May, Mission #183 hit Tokyo once more, with 502 B-29s attacking just north of Mission #181's target area, but nearer to the Imperial Palace. The new target included the familiar combination of factories and homes, and parts of the financial, commercial, and governmental districts of southern Tokyo, along with the Shinagawa railroad marshaling yards that connected Tokyo, Yokohama, and Kawasaki with the rest of Kyushu. Destruction of the Shinagawa yard would cripple ground transport all over the island.[171]

Cloud cover was light compared to Mission #181, but cloud and smoke forced most bombardiers to drop their 3,262 tons of incendiaries by radar. The defense was stiff; crews reported about eighty fighters and the heaviest, well-aimed flak of the campaign, and the toll of losses from all causes amounted to twenty-six B-29s destroyed and over a hundred damaged. At 5.2% of the sortied aircraft destroyed and nearly 25% damaged, the last Tokyo fire raid was the most costly of the war to XXI Bomber Command...and reminiscent of a bad day over Europe.

Photos showed the fires destroyed another 16.8 square miles, the greatest area wiped out in any single Tokyo raid, though Mission #40 had accomplished almost as much with about half the bomb weight and causing orders of magnitude more casualties. This time the firebombs damaged the Imperial Palace itself, destroyed the Empress Dowager's and the imperial prince's palaces, burned down the Navy Ministry, destroyed another 130,000 homes, and made another half million people homeless while killing only 700 and injuring another two thousand. The six Tokyo incendiary missions had burned out over fifty-six square miles—50.8% of the entire city area—slightly more than the sum of the actual designated target areas, which was roughly equal to two and a half

---

[*] Ugaki believed there were 400.

times the area of Manhattan. As they had with Nagoya, planners now scratched Tokyo from the list of incendiary targets.

Twentieth Air Force would not hit Tokyo with a major raid again: there were no more industrial targets there left to burn. Tokyo was never a nuclear target because there was practically nothing left to destroy after May 1945. With over 700,000 homes and over a million other structures destroyed, the Japanese people, and even their government, abandoned Tokyo to its fate. Nearly five million—roughly half the prewar population—either evacuated or just ran away. It was after this final fire attack on Tokyo that the Japanese high command got word[*] that XXI Bomber Command had *officially* changed their tactics to area firebombing. Ugaki, a keen if myopic observer, wrote that he had written off his home in Tokyo. It may have already been destroyed by the time he wrote that entry.[172]

Katsumoto Saotome joined the destitute after a raid destroyed most of his family's Mukojima ward home. Katsumoto's father collected sheet metal, while the boy scavenged for sardine cans in a ruined factory. "I was like a stray dog," Katsumoto said, "picking whatever looked valuable on the ground." Playwright Kafu Nagai, forced out of three homes by fire raids, chronicled his ordeal in his diary. "Every day I burn scraps of wood taken from destroyed houses to cook my meals," he wrote. "Life in a defeated country—no water, no fire. One can fairly say that we have reached the extreme of misery."[173]

Yet, even after a month and a half of firebombing, with millions of homes destroyed and millions of people scrounging to maintain even a mere subsistence level of life, the *official* Japanese reaction was, at the very best...silence. And no one in Japan seemed to notice:

> Japanese ideological mobilization and control was such that there are no signs of resistance to the government's suicidal perpetuation of the war at any time during the bombing campaign. Whatever the suffering, most Japanese then and subsequently appear to have accepted the legitimacy of the decision to continue fighting a hopeless war...[174]

But this was not *fighting* for the typical Japanese: it was *enduring*. Japan's leaders saw the incineration of their cities as the *expected result* of the nations' failure...

---

[*] A report from a source on Guam, according to Ugaki.

Yokohama

LeMay scheduled a daylight fire raid on Yokohama for 29 May (Mission #186). Situated on the crowded west shore of Tokyo Bay, separated from the capital by Kawasaki, Yokohama was Japan's fifth largest city and her second largest port, with a prewar population just shy of a million souls. It was an important shipbuilding and automotive center with factories devoted to those industries, as well as oil refineries, alumina processing plants, and chemical manufacturers. Yokohama suffered from spillovers during multiple raids, with the Tokyo raid on 15 April inflicting severe damage. However, planners had never before designated it as the primary target.

Relatively heavy losses in the recent Tokyo night missions were of some concern to both Arnold and LeMay. To avoid more serious casualties to interceptors by day, Mission #186 called for 517 B-29s in a high-altitude formation attack, with the groups crossing a time-control point on the Honshu coast at four-minute intervals. They assigned aiming points to each wing according to a schedule of downwind bombing, calculated to give the crews at least one drop in the target zone before smoke obscured it. Defending against the swarms of day fighters concentrated in the Tokyo Bay area, LeMay called on VII Fighter Command's Iwo Jima-based P-51s for their first assignment[*] as escorts on an incendiary mission.[175†]

The 454 B-29s that reached Japan found better weather than was usual in day missions, dropping 2,570 tons of M47 and M69 incendiaries. Over a hundred P-51s from VII Fighter Command joined the bomber stream at Fujiyama, that then split into separate flights to escort the various attack elements. Flying parallel, in trail, and about 2,000 feet above, the fighters met approximately 150 aggressive Japanese fighters. Early bomber formations bombed visually, but as was usual in such heavy attacks, smoke forced latecomers to resort to radar. The Mustang pilots claimed to have shot down twenty-six interceptors, claimed twenty-three more probables and damaged nine more, for a loss of only three of their number. Even with this help, the Japanese fighters and flak damaged 175 B-29s and downed five. Fires burned out the main business district along the waterfront, destroying an area of 6.9 square miles, a third of the city. The strike destroyed or damaged twenty

---

[*] One oral history source has the first escort joining the bombers on the Yokohama mission.

[†] P-51s out of Iwo Jima escorted *conventional* bombing missions as early as 7 April.

numbered targets. The total burned-out area now amounted to 8.9 square miles, somewhat more than the planned target area.[176]

On 30 May, O'Donnell held a 5-minute Memorial Day service on Saipan that included an invocation, O'Donnell's address:

Since the beginning of operations against the Japanese homeland six months ago, over one thousand of our officers and men, having taken off from this field on combat missions, failed to return, One thousand of our own is the terrible cost exacted thus far in the performance of our mission. That our mission has been well performed is no secret. That we have never faltered, despite our losses, in an undeniable truth. That the damage done to the enemy is all out of proportion to the cost is evident—but they are one thousand, they are our own. We miss them. We grieve for their families and loved ones, and today we humbly salute them.

Then…Taps.[177]

Osaka Again

With the principal urban areas around Tokyo Bay reduced to cinders, XXI Bomber Command turned westward to the urban complexes crowding the shoreline of Osaka Bay. There, in four closely spaced missions, LeMay brought the first phase of his incendiary program to a blazing end.

Even after Mission #42 on 13 March damaged Osaka, there were still untouched heavy and light industrial areas mixed in with port facilities, warehouses, storage dumps, shipbuilding yards, and petroleum installations lying along the Yodo River. Night tactics faced challenges in hitting these scattered areas since the core of the city had already burned out.

Thus, Mission #187 on 1 June was a daylight mission, launching 521 B-29s. Bomb loadings varied according to the specific targets assigned each wing, but every plane carried a 400-pound M28 (T4E4) fragmentation cluster bomb staged to fall before their incendiaries to discourage firefighters. Bombing from altitudes of 18,000 to 28,500 feet, 458 B-29s[*] dropped 2,788 tons on Osaka. Ton for ton, the attack was less successful than Mission #42—as expected—but the results still were

---

[*] Ugaki counted 350.

significant: 3.15 square miles burned out (plus a small area in nearby Amagasaki); over 136,000 houses and over 4,000 factories destroyed; nearly 4,000 dead or missing and nearly a quarter million *more* rendered homeless, and more numbered industries wiped off the map. Ten B-29s went down to all causes.[178]

Another 400,000 or more fled the city, but there were already so many refugees in the country that food was scarce, forcing many into the mountains to subsist on boiled pine needles and caterpillars.

Once again, VII Fighter Command was called on for escort, but the 148 P-51s launched from Iwo Jima had weather trouble en route to the rendezvous. The Mustang was a hard plane to fly in heavy weather with the required ventral fuel tank, and only twenty-seven Mustangs escorted the B-29s over the target, where the enemy reaction was stiff.[179]

Kobe Again

XXI Bomber Command had fire-bombed Kobe on the west shore of the bay on Mission #43, and spillovers from a precision attack against the Kawanishi aircraft plant in suburban Mikage on May 11 damaged it even further. But important districts to the east and west of the area razed by Mission #43 remained untouched. These contained business and residential sections, heavy and light industries, and transportation facilities.

On 5 June, Mission #188 struck Kobe with 473 B-29s (of 531 launched). As at Osaka, the heavily concentrated bombers were over the city for only an hour and a half as compared to the earlier three-hour raids. They went in unusually low for a daylight strike, dropping 3,077 tons of bombs from altitudes of from 13,650 to 18,000 feet. Enemy fighters were up in force to meet the unescorted bombers, making 647 attacks. Determined Japanese fighters and flak took down 11 B-29s, damaging 176, and operational causes took two more.

The attack eliminated Kobe as an incendiary target, razing 4.35 square miles and over 50,000 buildings, heavily damaging nearly another thousand structures. Aircrews saw columns of smoke rising 30,000 feet.[180]

Visible from everywhere on the island, survivors and refugees may have looked back at the towers of smoke, as Lot's wife may have looked back at Gomorrah. Still, their leaders did not wonder at their sin; they knew what it was.

Osaka Twice More

On 7 June it was Osaka's turn again. XXI Bomber Command's Mission #189 struck industrial and transportation targets and the east-central section of the city. Three wings carried incendiaries, but planes of the 58th Wing (from XX Bomber Command, recently redeployed from India) loaded 1,000-pound high explosives to drop on the Osaka Army Arsenal, Japan's largest arsenal and a prime source for IJA ordnance. Of the 458 Superforts launched, 409 dropped their bombs by radar because of a heavy undercast. Despite this and a moderately high attack altitude (17,900 to 23,150 feet), their 2,540 tons of bombs burned out 2.21 square miles and destroyed over 55,000 buildings, of which over a thousand were industrial. Fighters destroyed no B-29s on Mission #189; flak hit eleven. Mechanical failures claimed only two Superforts.

Once again, streams of refugees fled the city as the dogged survivors who stayed behind buried and burned the dead, scrounged for food and shelter...and waited. Their leaders, who the people could not choose, kept telling them to stay in their factories and fields. But as food and everything else grew shorter and shorter, how many of them wondered...why? Surprisingly—or perhaps not—none of the survivor accounts ever posed that question.

On 15 June, Mission #203 sent LeMay's Superfortresses back to Osaka to mop up. It was a token of the efficiency of the command's maintenance system that it was able, after a month of maximum-effort missions, to put up 516 B-29s, almost as many as had gone out in the first four-wing attack.

There were not enough landmarks left in Osaka to justify a maximum effort, so LeMay planned the mission with two mean points of impact[*] in Osaka and three in neighboring Amagasaki. This industrial suburb contained the Kawanishi aircraft factory, some large synthetic oil refineries, strategically important power plants, and many miscellaneous industrial firms.

This was not an example of the law of diminishing returns. Mission #42's fires swept through Osaka's most congested residential and commercial districts, creating new and extensive firebreaks that divided the remnants of the city into separate target areas. While the flames accounted for more than half of the total of 15.6 square miles consumed

---

[*] The point where it was desirable for the most bombs to fall.

in that attack and those that followed, they left 90% of the industrial roof area intact.

Over the two cities for two hours and eleven minutes, 444 B-29's dropped 3,157 tons of incendiaries. Because of the scattered nature of the targets, the total area burned over was less extensive than usual—an additional 1.9 square miles in Osaka and 0.59 square miles in Amagasaki, at a cost of two B-29s. Osaka was the last *major* firebomb target. The June raids, less impressive in total acreage consumed, hit industrial buildings more severely, accounting for an additional 25% of the roof area, though Ugaki called it "fairly heavy damage" in his diary.[181]

Mission #203 completed Phase I of the Twentieth Air Force's urban area program. LeMay accomplished the planners' objectives with dispatch and a thoroughness that was rarely possible in a comprehensive strategic bombardment plan. LeMay had departed from the schedule by scratching Yawata on the island of Kyushu and substituting Kobe, thus concentrating his efforts on three areas on Honshu.

In results, the campaign against the Japanese cities was more final than the Big Week had been in Europe.

By 11 June, LeMay's bombers had incinerated over one hundred square miles of Tokyo, Osaka, Nagoya, Yokohama, Kobe, and Kawasaki. Under the strain of successive major attacks, civil defense organizations collapsed, throwing an additional burden on the already overtaxed government. Nurtured on victory propaganda, under the stress of urban attacks, Japanese civilians showed little of the discipline which had carried British and German citizens through several years of aerial bombardment. Inhabitants of the great cities, stunned by the loss of the Marianas the previous summer, and now suffering the *effects* of that loss, lost confidence in their leaders' ability to defend them from attack or to care for the victims…but *did not rebel*. Instead, a quarter of Japan's urban population would *quietly* abandon the ashes of their cities. Abe Motoki, Minister of Home Affairs, said later:

> I believe that after the 23-24 May [sic] 1945 raids on Tokyo, civilian defense measures in that city, as well as other parts of Japan, were considered a futile effort.

As for XXI Bomber Command's cost, that had not been as dear. In seventeen maximum-effort incendiary attacks, LeMay had sortied 6,960 B-29s, carrying 41,592 tons of bombs. Losses had amounted to 136 aircraft, an average of 1.9 %, a rate well under that suffered during earlier months, and wholly acceptable according to conventional norms of strategic bombardment.[182]

While civilian casualties were catastrophic, the fire raids destroyed civilian habitats at a rate well beyond their ability to replace them. In Osaka alone, over a million people were homeless by the end of June. Whether we call it "terror bombing" or "strategic target destruction," the B-29s firebombing raids achieved widespread damage to Japan's infrastructure.[183]

And that was the mission Arnold created the Twentieth Air Force to perform.

Now, the smaller cities of Japan would burn…from one end of the archipelago to the other.

The Little Fire Blitz

Our parents would just say, 'That's a different era,' ... They wouldn't talk about it. And I figured my own family wouldn't understand.

Nihei Haruyo

On many days and nights that spring and summer, up to 5,500 B-29 crewmen commuted along the "Hirohito Highway" hauling bombs and mines, filling the skies over Japan with storm clouds of aluminum, rubber, and napalm. Back in the Marianas, ground crews hustled to swap out engines and patch bullet and flak holes to get more airplanes aloft. LeMay showed the potential of the B-29 and strategic air power, realizing the dream Arnold had set in motion years earlier when he fought tooth-and-nail for the Three-Billion-Dollar Gamble.[184]

On 16 June, the day after the last incendiary attack against the major cities, LeMay alerted his wing commanders of a new program that would begin on the following night: *single*-wing attacks, using the same tactics as before, but restricted to cities with populations over 50,000 but less than half a million. These towns were relatively congested, had war industries, and were transportation centers. The efficiency and light cost of the night raids and the weather outlook were convincing arguments. A list of 25 cities, with populations ranging from 323,200 (Fukuoka) to 62,000 (Hachioji) was tentative, but served well enough to get the campaign under way. This approach stressed the importance of cumulative effects of raids compressed within a short period that some felt might create a *possibility* of achieving a decisive effect with air power.

With few exceptions, Little Fire Blitz missions were like those against the major cities, but smaller in scale. Pathfinders marked the targets while the planes loaded the familiar combination of M47 and M69 incendiaries attacked using radar at altitudes between 7,000 and 9,200 feet. Enemy opposition, expected to be weak, was almost nil.* Missions #207-210 suffered one loss, and that was to unknown causes. The ratio of effective sorties for the 17-18 June missions, 456 out of 477, was high and the total weight of bombs dropped, 3,058 tons, was heavy. Omuta (Mission #207) received the heaviest attack from the 58th Wing of XX Bomber Command but suffered least, with a destroyed area of less than a quarter of a square mile, only 4.1% of the city's area. For Mission #208 on Kagoshima, parts of the 73rd Wing burned 2.15 square miles, or 44.1%. Parts of the 73rd Wing burned 2.44 square miles, or 70%, of Hamamatsu in Mission #209. At Yokkaichi (where only three 313th Wing groups bombed), Mission #210 burned 1.23 square miles, or 60%. The total area burned out, 6.037 square miles, was considerably better than the average results achieved by four-wing missions against a single city, but destroyed some 80,000 homes and left another 80,000 people homeless. Ugaki wrote, "[b]efore dawn, the western sky looked like the evening glow."[185]

These "smaller" cities were still urban areas that provided services to the surrounding countryside, the farmers and fishermen, the country doctors, and the smaller shops where craftsmen sold their wares. Until that summer, most had lived undisturbed by the war except to send their sons off to fight. That summer, the B-29s burned the cities while US Navy submarines sank the fishing boats, having destroyed all their other targets. More refugees streamed into the already overtaxed countryside to scrounge for what they could find. Local firemen, like those in the big cities, shouldered their flails and shovels and shuffled past the fields of ash dotted with ruined machine tools on their way to their posts.

The success of the first multiple-target mission insured the continuance of the program, and the operational pattern established on 17-18 June became standard for the rest of the war. Whenever a force of B-29s was ready to go with radar conditions predicted, XXI Bomber

---

* As the Okinawa campaign went on, Japan started husbanding their aircraft and pilots for the final battle of the Home Islands, where they anticipated using all their planes as kamikazes.

Command scheduled a night incendiary mission. Targets for a particular night were based on operational considerations—weather, radar conditions, and relative position of the several towns—as well as on data furnished by intelligence reports. As the campaign progressed, the targets became smaller and of less significance, making it harder to accurately calculate the type and weight of bombs required for any single mission.

On strike nights, XXI Bomber Command usually attacked four cities, with one wing assigned to each. Occasionally, target cities were large enough to require two wings. Two such instances included Mission #211, Fukuoka/Miyazaki, on 19 June that destroyed eight thousand houses, and Mission #294, Omuta/Tokuyama, on 26 July. Sometimes, planners integrated the 315th Wing's night strikes against petroleum targets with incendiary attacks, and coordinated day and night missions in a furious round-the-clock effort until the end of the war.

LeMay sent multiple incendiary attacks on sixteen nights that summer, an average of two a week, directing the main weight of attacks against urban areas, as he had been doing since March. Between 9 March and 14 August, about 70% of the bombs dropped were in incendiary attacks. After 17 June, 8,014 sorties dropped 54,184 tons of incendiaries on 58 secondary cities. As the planners scratched original targets and added others, by 14 August 58 towns had been fire-bombed.

On Mission #308 on 1-2 August, the B-29s of the 313th Wing destroyed 97% of the city of Toyama, an aluminum and ball bearing production center of some 150,000 people, about the size of Chattanooga, while simultaneously burning out five other nearby urban areas—with just 43 aircraft. By the end of the war, incendiary bombing would consume 178 square miles of 69 Japanese cities.[186]

Arnold to LeMay: How Long?

The bombers ruined the six most important industrial cities in Japan and destroyed or damaged nearly all the great factories. The bombers incinerated thousands of household and feeder factories and shops. Casualty lists ran into six figures, approaching *seven* figures when factoring in starvation that inevitably followed. Millions had lost their homes, and the evacuation of survivors to the country, where food was already scarce, made it hard to secure labor for the few factories that remained.

In all the cities, the area attacks worked vast hardships upon the Japanese people. Statistics of dead, wounded, and homeless tell little about the personal sufferings, nor about the dislocation which occurred as thousands fled the towns to take their chances in the countryside. The effect of this dispersal on industry, as expected, was great. So also was the effect on morale. In a report on the effects of urban area bombing submitted in December 1945, a group from the faculty of the Imperial University of Tokyo wrote that:

> ...with the shifting of the attacks from cities to local districts, the people became concerned over the future of the war. In consequence, their fighting morale was weakened.[187]

In mid-June, Arnold flew out to Saipan to confer with LeMay about the campaign, and asked how long before XXI Bomber Command would be out of targets. LeMay's analysts said mid-October.

After the Tokyo fire raids of 23 and 25 May, Joseph C. Grew, Ambassador to Japan until Pearl Harbor and Acting Secretary of State in the spring of 1945, attempted to persuade Harry Truman to temper his unconditional surrender message of 8 May with a statement that the United States had no intention of abolishing the emperor's office. Grew believed that with such a guarantee, the Japanese might be willing to capitulate. Truman was sympathetic, but feared that the Japanese might interpret such a concession during the Okinawa fighting as a confession of weakness. For the rest of his life, Grew maintained his belief that the war might have ended earlier if Truman had made that offer.

Grew may have known Japanese politicians and diplomats, but neither he nor any other outsider had any idea about what was going on inside Japan's 1945 halls of power. While there was a sharp slump in Japanese civilian morale in the wake of the Fire Blitz—borne out by postwar opinion surveys—plus a renewed effort by some minor officials outside Japan to negotiate for peace, there is no evidence that most of the senior generals and admirals were interested in peace at all; Ugaki makes no mention of any peace overtures or even discussions in his diary.

While a few senior Japanese officers and politicians might have been quietly discussing ways to end the slaughter, there were junior officers who stalked those same halls ready to kill—literally—anyone who even *thought* about capitulation. In the *bushido*-infused minds of these "young men of purpose" known as *shishi*, the destruction of all Japanese—and Japan—was the only *possible* consequence of Japan's failure to achieve the resource autarky that the military leadership had set for them. To the samurai of only a century before, failure to complete any task, large or small, was not *punishable* by death; *death was the only logical consequence of any failure.*

All that summer, the Americans implored Japanese leaders to at least respond to repeated radio messages. Japanese officials met these entreaties with silence.

How long *would* Japan hold out? Would the "victors" of this drawn-out conflict have to occupy a graveyard?

The Emperor's Voice

By mid-August, electricity only worked for an hour or two at most in nearly all of Japan. Furthermore, the militarists did not fulfill *any* of the glorious promises they made before or during the war.

On 14 August, Japan Broadcasting Corporation (NHK) technicians set up microphones in an office bunker under the Imperial Household Ministry and recorded two versions of the Showa Emperor Hirohito's reading of his Imperial Rescript on the Termination of the War. NHK would broadcast the recording the next day.

The Americans, British, Soviets and Chinese had killed more Japanese men between 1932 and 1945 than had been born in the half-decade before. Within the nine days prior to the broadcast, two cities that had been largely untouched by the war all but vanished in the blink of an eye.

But...

Elements of the IJA refused to accept that the Emperor was going to end the war. In what would become known as the Kyujo Incident, scores—some say hundreds—of soldiers raided the Imperial Palace that evening to destroy the recordings, battling with the Imperial Guard. The attempted coup failed; NHK officials had already smuggled the recordings to the studio. Major Hatanaka Kenji attempted to block the broadcast by occupying the NHK station, but the Eastern District Army ordered him to desist; he later committed suicide.

On 15 August,[*] all NHK stations announced the Emperor would address the nation at noon. This was a momentous event: it would be the first time that common Japanese had heard the voice of *any* Japanese

---

[*] 14 August in the US.

Emperor and the first radio address by the Emperor. Many wore formal clothes for the occasion.

At precisely noon, an NHK announcer instructed the nation to stand for an announcement "of the highest importance." The stations played *Kimigayo*, followed by the Emperor's rescript.

Though the Showa did not explicitly *state* the word "surrender," he instructed his Prime Minister Suzuki Kantaro and his administration to tell the Allies that the "Empire accepts the provisions of their joint [Potsdam] declaration" of the United States, the United Kingdom, China, and the Soviet Union. He justified Japan's decision to go to war as an act of "self-preservation and the stabilization of East Asia" and referenced the setbacks and defeats of recent years, saying "the war situation has developed not necessarily to Japan's advantage." He mentioned the atomic bombs obliquely, calling them a "new and most cruel bomb." The Emperor ended with a call on the Japanese people "to be devoted to construction for the future."

He did *not* mention the fire bombings.

The Showa delivered the address in formal, Classical Japanese, using pronunciation and intonations unfamiliar to ordinary Japanese. Because the speech made no direct reference to *surrender*, it confused listeners not familiar with the Potsdam Declaration—which would have been nearly all of them—whether Japan *had* actually surrendered. The poor quality of the recording, of the radio broadcast, and the Showa's formal courtly language worsened the confusion.

This reaction had been expected. A radio announcer familiar to most Japanese listeners came on the air immediately after the recording ended to explain that the Emperor had, indeed, meant that Japan *was* surrendering. Robert Guillain wrote that most Japanese retreated to their homes or places of business for several hours to quietly absorb and contemplate the significance of the announcement.

In just the last six months of the war, LeMay and his B-29s had incinerated over fifty Japanese cities, large and small. Oil production and distribution was nearly nil. Mines laid by his bombers had brought Japanese fisheries and commerce to a halt. The fire raids killed nearly a million Japanese, injured three times as many, and had driven millions more from their homes.

LeMay and the B-29s ensured that the Japanese people, contemplation or not, would have absorbed enough.

Vindication, Vengeance and Marketing

*Of what use is decisive victory in battle if
we bleed to death as a result of it?*

B.H. Liddell Hart

An Enemy No One Foresaw

Was *this* the *kind* of war that Douhet and Mitchell had envisioned? Wasn't it closer to the kind that the Americans so deplored when Arthur Harris advocated area bombing against Germany—his infamous de-housing program? The prewar theorists firmly believed that their strategies would, at the very least, scare the enemy leadership into either surrendering or entering negotiations. But the utter desolation wrought by WWII would have struck Mitchell and Douhet's 19th century sensibilities as revolting...just as the reality of what the Americans *needed* to defeat Japan revolted Hansell.

The theorists of the ACTS conceived air warfare not in a vacuum, but in the rarefied atmosphere of the *end* of a horrid conflict that few understood—World War One. That horror was very much a 20th Century war and *not*, as some have claimed, the last war of the 19th Century. It *was* a global 20th Century conflict but, perhaps, it *ended* in a *very* 19th

Century, *very* European manner. The two most powerful European victors (Britain and France) wrote the Treaty of Versailles (officially, the Treaty of Peace Between the Allied and Associated Powers and Germany) and held it out for their humbled former enemy to sign. And sign they did, if under protest. Minor allies and the Associated Power (the US) watched as Britain and France charged Germany for millions of deaths and devastation that took generations to repair—and still haven't repaired. A series of conferences and banquets dividing up the spoils followed, where the minor allies got what they could for their troubles. Belgium, Portugal, and Italy got a pittance; the Americans wanted nothing (but *got* aspirin); Japan got most of Germany's western Pacific holdings that included the Mariana Islands north of the equator, the biggest being Saipan and Tinian.

The dictatorships of Germany and Japan in 1945 would have been completely impossible by the lights of 1918. Douhet and Mitchell could not have imagined such states could exist.

The kind of society that *they* expected to cow into submission died during the Great War, though few even *sensed* it. Similarly, though the beliefs of WWII air power enthusiasts—based on a 1918 mindset—were sincere and well-meaning, they did *not* envision the need for long-term occupation, nor for war crimes trials resulting in executions. They could not imagine the enduring strength of a monomaniacal and uncaring German dictator who felt that his country was unworthy of him, or of die-hard Japanese medievalists willing to fly their airplanes into ships and see their entire country immolated for *their* failure to achieve the goals they set out for them.

The nature of Japan's war with the West, goaded by the IJA and started by the IJN, is best summed up by Commander Okumiya Masatake, who wrote,

The Pacific War was started by men who did not understand the sea and fought by men who did not understand the air.[188]

Perhaps such vastly destructive area bombardment *was* required by the nature not just of the targets themselves, but the nature of the opponents. Dwight D. Eisenhower's classic memoir, *Crusade in Europe*, is more than just reminiscences of the Supreme Commander. It is a confirmation of what some commentators have suggested: that Central and Eastern Europe, despite centuries of economic and cultural engagement with the West, *still* entertained the idea of eradicating millions—simply because of their faith—as a military necessity.

Similarly, the dominant samurai of 20th Century Japan still believed they could resolve all problems through combat. Japanese society—as a whole—had not reached a level of civilization where their leaders would find the horrors of the Nanking massacres utterly repulsive.

It was these two cultures that drove WWII to the brutal depths of depravity that included the death camps of Europe and the systematic murder of the Chinese. The response to these horrors were the de-housing campaign in Europe, and the Fire Blitz in Japan.

The Power of Postwar Marketing

Though LeMay argued during and after the war that it was industry that was targeted by the Fire Blitz, the USSBS was honestly ambiguous. Under the heading "Effect of Target Selection Theory on the Air Campaign against Japan Proper," *Report #53 (Pacific)* states:

> The purpose of the strategic bombing campaign was, until the last weeks of the war, the reduction of the armed strength with which the enemy could be expected to oppose a landing by our ground troops in November 1945.[189]

This statement of purpose highlighted what the Fire Blitz planners intended: to avoid or mitigate the need for opposed invasion. Even if bombing destroyed much of the Japanese industry, the USSBS believed that:

> Had attention been directed towards spectacular increases of pressure rather than toward facilitating invasion, the railroads would have presented a promising target. Railroad attack would have carried with it an almost immediate threat of starvation, not only for the major urban concentration but for the entire deficit food areas, such as western Honshu, and would have placed a severe restriction on internal communications. In view of the already overstrained truck, motor fuel, and shipping positions, the dependence of the railroads themselves on the continuance of coal traffic over only four lines, and the importance of food shipments from Hokkaido, there is very good reason to believe that an effective railroad attack might have brought about a very rapid capitulation.[190]

While true, there's nothing in the USSBS about the millions of people who fled the cities during the war, starting in March 1945...*after* the Fire Blitz had begun. Japan's poverty already crippled their industry, and losing millions of workers could not have helped.[191]

Ultimately, the war that began so suddenly for the United States ended by changing only *one* mind: that of the Showa Emperor Hirohito. Because Japan's military leadership was dead-set on their own immolation if need be, the spiritual leader of Japan—in whose name the military justified every act of aggression right up to August 1945—had to tell them to stop. He—as his grandfather, the Meiji Emperor, had done before him—had to *briefly* exercise more temporal power than any Japanese emperor ever had before...or ever will again.

In his papers, he stated he could only bring himself to "bear the unbearable" reality of defeat when he was told that American and Soviet landings were inevitable. And *that* revelation came *between* the attacks on Hiroshima and Nagasaki, albeit after the Soviet Union declared war on Japan. While certainly important, the fire raids were *not the* deciding factor, but most certainly were several of many. Indeed, it is likely that the Showa—his thinking *spurred* by the fire raids—simply decided that with everything *else* that had happened, he and his country had had enough. Although *not* raised in samurai traditions—yet fully aware of them—he chose to save lives rather than sacrifice them to an ancient and flexible[*] concept of "honor."[192]

So, for that matter, did Hansell, with his insistence on precision daylight bombing. But *his* way wasn't working in Japan, so LeMay took over, with the vision and flexibility to change USAAF methods and make the B-29s effective.

In retrospect, we can compare and contrast the genteel patrician Hansell and the hardscrabble LeMay to the disparity between patrician Robert E. Lee and plebeian Ulysses S. Grant. Lee, the patrician creature of the agrarian South; and Ulysses S. Grant, the plebeian man of the Industrial Revolution North. Lee fought the way he knew how with the tools he had; Grant did the same, but he invented new ways of war as he went along. Lee's way worked sometimes; Grant's worked *all* the time.

Discounting their miscalculation of Soviet resistance to Germany, Japan intended for the death and destruction they wrought in the early months of the war to scare the West into peace. It was a tremendous

---

[*] *Bushido* has been defined by many commentators in different ways over the centuries. Its definitions of what *was* honorable changed several times.

gamble that failed completely. America's first low-level incendiary attack on Tokyo had, too, been a gamble, but unlike Pearl Harbor, the fire raids could be—and *were*—repeated. The six month-long Fire Blitz showed the USAAF planners—but *convinced few* others—that a costly amphibious invasion of the Home Islands was unnecessary. The senior leadership in the Bomber Mafia, who also knew of the atomic bomb being developed, was confident that their current aircraft and techniques could bring Japan to defeat.

That, *and…*

The Showa's radio address spoke not of the sea mining, or the fire raids, or losing islands or shipping, or the impending Soviet invasion. The only weapons he mentioned specifically were the atomic bombs, delivered by the B-29s of the Twentieth Air Force.

There is no proof—nor can there be—that the destruction of Japanese urban areas and the ensuing psychological damage *negated* the need for invasion. Scholars have made a case, and the Showa himself provided evidence that the atomic bombings were only the last straws on an exhausted camel's back.

But the air power enthusiasts *believed* they had done it all themselves, and have since convinced many others—via the Strategic Bombing Survey and many other documents, films and other mass media that were expressly created to support the idea of an independent air arm and of the USAAF's premier role in ending the war—even if that message is sometimes ambiguous.

They achieved their goal with the creation of the US Air Force in 1948.

For that marketing job alone, the shades of Douhet and Mitchell might have smiled.

Bibliographic Note

The end of WWII has always been a matter of heated debate, with the greatest controversies arising over what *actually* ended it. My crude efforts here have been an attempt to bring clarity to the US Air Force scenario—the single-cause theory—which they've been noisily fostering since Nagasaki. I do not subscribe to this theory that the USAF has pushed since 1945.

The Allies beat Germany into the ground; so did they Japan, but "Japan" had no intention of giving up because of its leadership's beliefs…save one. That story is long and is better explored in the book I wrote with Lee Rochwerger, *Why The Samurai Lost Japan: A Study of Folly and Miscalculation.*

The story of air power before WWII has, like many other nebulous subjects, developed and changed after every major US conflict. Walter Boyne's *Influence of Air Power Upon History* has the advantage of only examining the results, not the evolutions themselves.

There have been several books on the Pacific bombing campaign, but few are definitive. The US Strategic Bombing Survey—several score volumes of it—is probably the best summation, but Curtis LeMay's *Phase Analysis* report is a rather dry summation of only the first ten days of the Fire Blitz. The USAAF's official history, edited by Wesley Craven, is again dry, but comprehensive.

There are scores of books on the B-29, many of them redundant, but there are only a few on aircrew experience. Robert F. Dorr's *Mission to Tokyo* has the most oral history accounts of men who undertook the Fire Blitz attacks.

Both LeMay and Hansell wrote memoirs of their war experiences, and both have their value, but like any other memoirs, they need to be looked at with a jaundiced eye and taken with more than a grain of salt. Other air commanders in the Marianas had their diaries published after the war. Aircrew diaries are unknown, but Dorr has extensive interviews.

Japanese oral histories are few and far between because in Japanese culture, unpleasant things are just not talked about. A researcher undertook the only known major endeavor to conduct an extensive set of interviews of survivors of the first Tokyo raid that appeared in the *Japan Times* in 2015. Most of the interviewed subjects were children in 1945. Other oral testimonies appear in the few other books written about the Fire Blitz and its aftermath.

Index

Boeing B-29, 9, See also B-29
Bomber Mafia, 21, 23, 112
bombsight, Speery/Norden, 37
Britain, 17, 18, 26, 27, 109
Burma, 47
bushido, 18, 105
Butte (MT) *Standard*, 74
Canfield, Robert, 61
Caproni, bomber, 19
Casablanca conference, 25
Catton, Jack, 61
Chattanooga, Tennessee, 104
Chengdu, China, 48
Chengtu Province, China, 25, 47
Chennault, Claire, 25
Chiang Kai-shek, 25
Chiba Peninsula, Japan, 65
China, 10, 25, 28, 40, 43, 47, 48, 49, 50, 62, 107
Chiyoda ward, Tokyo, Japan, 93
Churchill, Winston, 25, 27
COA, Committee of Operation Analysts, 30, 31, 39, 49, 50, 53, 56
Cologne, Germany, 29, 41
combat box, 24, 36
Command and General Staff College, US, 20
Committee of Operations Analysts (COA), 30
Copeland, Robert, 85
critical nodes, 29, 39
Curtiss, Glenn, 19
Davies, John, 63
Directorate of Bombardment, USAAF, 31
Dolittle, James H. "Jimmy", 13, 14, 40, 42, 44, 90
Douhet, Giulio, 16, 18, 19, 20, 21, 26, 27, 108, 109, 112
Dublon Island, Truk Attoll, 53
Earthquake Memorial Hall, Tokyo, Japan, 73
Eighth Air Force, US, 24, 30, 90

Eisenhower, Dwight D., 109
Europe, 16, 18, 22, 23, 24, 26, 27, 29, 31, 34, 36, 38, 39, 41, 45, 50, 54, 58, 60, 70, 87, 94, 100, 109, 110
FD-2 airborne radar, 45
firebreaks, 39, 41, 77, 80, 83, 99
Fourteenth Air Force, US, 25
France, 17, 19, 22, 26, 109
Fukagawa ward, Tokyo, Japan, 73, 86
Fukuoka, Japan, 102, 104
Funato Kazuyo, 66, 71
Georgia, state, 35
Germany, 17, 18, 26, 29, 51, 74, 108, 109, 111, 113
Gobunko bunker complex, Imperoal Palace, Tokyo, Japan, 71
GOMORRAH, Operation, 56
Grant, Ulysses S., 111
Great Bend, Kansas, 35
Great Kanto Earthquake, 1 Sep 1923, 30, 82
Grew, Joseph C., 105
Guam, Mariana Islands, 51, 52, 57, 60, 61, 63, 91, 95
Guillain, Robert, 69, 107, 121
H2X radar, 38
Hachimanga Shrine, Tokyo, Japan, 86
Hachioji, Japan, 102
Hamacho, Nihonbashi, Tokyo, Japan, 64
Hamamatsu, Japan, 103
Hamburg, Germany, 29, 41, 56, 67
Hankow (Wuhan), China, 49, 50
Hansell, Hayward S., 23, 24, 26, 27, 30, 54, 55, 56, 57, 58, 59, 88, 108, 111, 114
Hatanaka Kenji, 106
Hawaii, 23, 24, 54
Haynes, Caleb, 24

Mariana Islands, 10, 25, 30, 40, 43, 50, 51, 52, 53, 54, 57, 58, 62, 90, 100, 102, 109, 114

Marshall, George C., 22, 25

MATTERHORN, Operation, 47

Maxwell Field, Alabama, 20

MEETINGHOUSE, Tokyo, Japan, 60

Meiji Theater, Tokyo, Japan, 69, 70

MICROSCOPE, Nagoya, Japan, 77, 85

Milwaukee (WI) *Sentinel*, 74

Mitchell, Williiam L., 19, 20, 21, 22, 23, 26, 27, 108, 109, 112

Miyazaki, Japan, 104

Musashino, Japan, 53, 55, 59

Nagoya Castle, Nagoya, Japan., 92, 93

Nagoya, Japan, 42, 44, 55, 56, 60, 77, 78, 79, 80, 81, 82, 85, 87, 92, 93, 95, 100

Nakajima J1N1 *Gekko*, Irving fighter, 45

Nakajima KI-44 *Shoki*, Tojo fighter, 45, 46

New Delhi, India, 47

New Orleans (LA) *Times-Picayune*, 81

Nihonbashi ward, Tokyo, Japan, 64, 73

Nitto Maru, 12

Norden bombsight, 39

Norden, Carl, 38

Norstad, Lauris, 24, 26, 55, 57, 58, 62, 88

North Africa, 24, 26

North Field, Guam, 63

O'Donnell, Emmet, 30, 53, 54, 56, 57, 63, 97

Ohtaki Masaharu, 69

Okinawa, 89, 90, 103, 105

Okumiya Masatake, 109

Omuta, Japan, 103, 104

Ono Kimie, 64, 65, 66, 69, 70

Operational range, aircraft, 34

Osaka, Japan, 42, 44, 56, 60, 68, 79, 80, 81, 82, 84, 85, 87, 97, 98, 99, 100, 101

Ostfriesland, German battleship, 20

Pacific Ocean, 9, 24, 25, 27, 28, 29, 31, 41, 46, 51, 55, 89, 109, 110, 113

Palembang, Netherlands East Indies, 43

Panama Canal, 23

PEACHBLOW, Osaka, Japan, 79

Pearl Harbor, Hawaii, 9, 22, 24, 28, 40, 105, 112, 121

Pellet, Marcian, 85

Pengshan, China, 48

Philippine Islands, 19, 22, 23, 25, 28, 87

Phillips, Charles, 64

Portugal, 109

Potsdam Declaration, 107

Powers, Robert, 63, 65, 109

Pratt, Kansas, 35

precision, 16, 20, 21, 23, 30, 36, 37, 55, 79, 83, 88, 90, 92, 93, 98, 111

Precision bombardment, 36

R-2600 Twin Cyclone, engine, 33

R-3350 Duplex Cyclone, engine, 33

RAF, 26, 29, 44, 45, 50, 56

Renton, Washington, 35

Robertson, Gordon B., 70

Roosevelt, Franklin D., 22, 25, 27

Royal Air Force (RAF), 26

Ryogoku Bridge, Tokyo, Japan, 72

Ryogokubashi ward, Tokyo, Japan, 66

Saigon, Vietnam, 71

End Notes

Preludes

[1] Harries, Susie, and Merion Harries. *Soldiers of the Sun: The Rise and Fall of the Imperial Japanese Army.* (New York: Random House, 1991), 396.

[2] Toland, John B. *The Rising Sun: The Decline and Fall of the Japanese Empire, 1936-1945.* (New York, NY: Bantam Books, 1971), 350; Ugaki Matatome, Donald M. Goldstein, Ed.. *Fading Victory: the Diary of Admiral Matatome Ugaki.*(Annapolis, MD, Naval Institute Press, 1991), 111.

[3] Toland, John B. *The Rising Sun: The Decline and Fall of the Japanese Empire, 1936-1945.* (New York, NY: Bantam Books, 1971), 350; Ugaki Matatome, Donald M. Goldstein, Ed.. *Fading Victory: the Diary of Admiral Matatome Ugaki.*(Annapolis, MD, Naval Institute Press, 1991), 112.

[4] Toland, John B. The Rising Sun: The Decline and Fall of the Japanese Empire, 1936-1945. (New York, NY: Bantam Books, 1971), 350-352.

[5] Guillain, Robert. *I Saw Tokyo Burning: An Eyewitness Narrative from Pearl Harbor to Hiroshima.* (William Byron, trans.)(New York, Ballentine Books, 1981), 59; Toland, John B. *The Rising Sun: The Decline and Fall of the Japanese Empire, 1936-1945.* (New York, NY: Bantam Books, 1971), 352.

[6] Scott, James M. *Black Snow: Curtis LeMay, the Firebombing of Tokyo, and the Road to the Atomic Bomb.* (New York, WW Norton & Company, 2022)(Kindle Edition), 229.

[7] Scott, James M. *Black Snow: Curtis LeMay, the Firebombing of Tokyo, and the Road to the Atomic Bomb.* (New York, WW Norton & Company, 2022)(Kindle Edition), 242.

[8] Scott, James M. *Black Snow: Curtis LeMay, the Firebombing of Tokyo, and the Road to the Atomic Bomb.* (New York, WW Norton & Company, 2022)(Kindle Edition), 178, 242.

The Leading Characters, Organizations, and Ideas

9 Douhet, Giulio. *The Command of the Air*. (Washington, D.C.: Office of Air Force History, 1983) (PDF Edition), 28.

[10] Kohn, Richard H, *Strategic Air Warfare*. (Washington, DC, Office of Air Force History, United States Air Force, 1988)(PDF Edition), 20

[11] Berger, Carl A. *B-29, The Superfortress*. (New York, Ballentine Books, 1970), 41-44, 45.

[12] Berger, Carl A. *B-29, The Superfortress*. (New York, Ballentine Books, 1970), 64; Dorr, Robert F, *B-29 Superfortress Units of World War 2*. (Oxford, England, Osprey Publishing, 2002) (PDF Edition), 15.

[13] Berger, Carl A. *B-29, The Superfortress*. (New York, Ballentine Books, 1970), 11-13, 17, 21.

[14] Kohn, Richard H, *Strategic Air Warfare*. (Washington, DC, Office of Air Force History, United States Air Force, 1988), 28.

15 Bond, Horatio, ed. Fire and the Air War; A Compilation of Expert Observations on Fires of the War Set by Incendiaries and the Atomic Bombs, Wartime Fire Fighting, and the Work of the Fire Protection Engineers Who Helped Plan and the Destruction of Enemy Cities and Industrial Plants. Boston, MA: National Fire Protection Association International, 1946, 139-140.

[16] Scott, James M. *Black Snow: Curtis LeMay, the Firebombing of Tokyo, and the Road to the Atomic Bomb*. (New York, WW Norton & Company, 2022)(Kindle Edition), 99.

[17] Bond, Horatio, ed. Fire and the Air War; A Compilation of Expert Observations on Fires of the War Set by Incendiaries and the Atomic Bombs, Wartime Fire Fighting, and the Work of the Fire Protection Engineers Who Helped Plan and the Destruction of Enemy Cities and Industrial Plants. (Boston, MA: National Fire Protection Association International), 1946, 138, 141; Bradley, F. J. No Strategic Targets Left. (Paducah, KY: Turner Pub. Co., 1999), 33; Cahill, William L, "Imaging the Empire: The 3rd Photographic Reconnaissance Squadron in World War II." Air Force History, Spring 2012, 14; Mann, Robert A. The B-29 Superfortress: a Comprehensive Registry of the Planes and Their Missions. (Jefferson, NC: McFarland, 2009), 140-1; Tillman, Barrett. "When Fire Rained from the Sky." Aviation

History, September 2016. http://www.historynet.com/ when-fire-rained-from-the-sky.htm, 34.

[18] Scott, James M. *Black Snow: Curtis LeMay, the Firebombing of Tokyo, and the Road to the Atomic Bomb.* (New York, WW Norton & Company, 2022)(Kindle Edition), 41.

[19] Unknown. History of the Organization and Operations of the Committee of Operations Analysts. (Washington, DC, USAAF, 1944)(PDF Edition), 1.

[20] Unknown. History of the Organization and Operations of the Committee of Operations Analysts. (Washington, DC, USAAF, 1944)(PDF Edition), 11.

[21] Unknown. History of the Organization and Operations of the Committee of Operations Analysts. (Washington, DC, USAAF, 1944)(PDF Edition), 20.

[22] Scott, James M. *Black Snow: Curtis LeMay, the Firebombing of Tokyo, and the Road to the Atomic Bomb.* (New York, WW Norton & Company, 2022)(Kindle Edition), 99.

The Tempestuous Superforts and their Target

[23] Berger, Carl A. *B-29, The Superfortress.* (New York, Ballentine Books, 1970), 27; Craven, Wesley F (ed). *United States Army Air Forces in World War II: Volume 5, The Pacific, Matterhorn to Nagasaki.* (Washington, DC, US Printing Office, 1983)(PDF Edition), 609; Dorr, Robert F, *B-29 Superfortress Units of World War 2.* (Oxford, England, Osprey Publishing, 2002)(PDF Edition), 6-7; Gorman, G Scott. *Endgame in the Pacific: Complexity, Strategy and the B-29: A Fairchild Paper.* (Maxwell AFB, AL: Air University Press, February 2000)(PDF Edition), 15; Harris, Samuel R. Jr. (Robert A. Mann, Ed). *B-29s Over Japan 1944-45: A Group Commander's Diary.* (Jefferson, NC, McFarland & Co. Inc., 2011); LeMay, Curtis E. and Bill Yenne. *Superfortress: The Story of the B-29 and American Air Power.* (New York: McGraw-Hill, 1988) (PDF Edition), 21.Nijboer, Donald. *B-29 Superfortress vs KI-44 "Tojo" (Duel).* (Oxford, England, Osprey Publishing, 2017 (E-book Version Bloomsbury Publishing)(Kindle Version), 3.

[24] The Story of The Billy Mitchell Group, 468 Heavy Bomb Group, From the C.B.I. to the Marianas. http://www.468thbombgroup.org, Accessed 18 Jan 2017.

[25] Kohn, Richard H, *Strategic Air Warfare.* (Washington, DC, Office of Air Force History, United States Air Force, 1988), 26; The Story of The Billy Mitchell Group, 468th Heavy Bomb Group, From the C.B.I.

to the Marianas. http://www.468thbombgroup.org. Accessed 18 January 2017.

26 Berger, Carl A. *B-29, The Superfortress.* (New York, Ballentine Books, 1970), 35; Phillips, Edward H. "Boeing's B-29: Birth of a Bomber." *Aviation History 8*, No. 5 (1998) (PDF Edition), 4-7; Stout, Wesley W. *Great Engines and Great Planes.* (Chrysler Corporation, Detroit, MI, 1947); The Story of The Billy Mitchell Group, 468 Heavy Bomb Group, From the C.B.I. to the Marianas. http://www.468thbombgroup.org. Accessed 18 Jan 2017.

[27] The Story of The Billy Mitchell Group, 468th Heavy Bomb Group, From the C.B.I. to the Marianas. http://www.468thbombgroup.org. Accessed 18 January 2017.

[28] The Story of The Billy Mitchell Group, 468th Heavy Bomb Group, From the C.B.I. to the Marianas. http://www.468thbombgroup.org. Accessed 18 January 2017.

[29] Town, Dick. "B-29 Flight Characteristics." (Bennington, VT, WW2 Journal, 1989), 11-12.

[30] Nijboer, Donald. *B-29 Superfortress vs KI-44 "Tojo" (Duel).* (Oxford, England, Osprey Publishing, 2017 (E-book Version Bloomsbury Publishing)(Kindle Version), 3-5.

[31] McFarland, Stephen L. *America's Pursuit of Precision Bombing: 1910-1945.* (Washington, DC: Smithsonian Institute Press, 1995), 197.

[32] Scott, James M. *Black Snow: Curtis LeMay, the Firebombing of Tokyo, and the Road to the Atomic Bomb.* (New York, WW Norton & Company, 2022)(Kindle Edition); Unknown, "Exhibit I Estimated Population and Labor Force 1944 for Selected Japanese Cities." Typescript (PDF), 4.

[33] Daniels, Gordon, "Before Hiroshima: The Bombing of Japan, 1944-45." *History Today*, January 1982, 15.

[34] Daniels, Gordon, "Before Hiroshima: The Bombing of Japan, 1944-45." *History Today*, January 1982, 15.

35 Bond, Horatio, ed. Fire and the Air War; A Compilation of Expert Observations on Fires of the War Set by Incendiaries and the Atomic Bombs, Wartime Fire Fighting, and the Work of the Fire Protection Engineers Who Helped Plan and the Destruction of Enemy Cities and

Industrial Plants. Boston, MA: National Fire Protection Association International, 1946, 138-39.

36 Bradley, F. J. *No Strategic Targets Left*. Paducah, KY: Turner Publishing Company, 1999, 34; Civilian Defense Division. United States Strategic Bombing Survey (Pacific): *Summary Report Covering Air Raid Protection and Allied Subjects in Japan (Report #10)*. Washington DC: U.S. Government Printing Office, 1946 (PDF Edition), 4-6; Daniels, Gordon, "Before Hiroshima: The Bombing of Japan, 1944-45." *History Today*, January 1982, 15.

37 Bond, Horatio, ed. Fire and the Air War; A Compilation of Expert Observations on Fires of the War Set by Incendiaries and the Atomic Bombs, Wartime Fire Fighting, and the Work of the Fire Protection Engineers Who Helped Plan and the Destruction of Enemy Cities and Industrial Plants. Boston, MA: National Fire Protection Association International, 1946, 138; Bradley, F. J. No Strategic Targets Left. Paducah, KY: Turner Publishing Company, 1999, 34; Civilian Defense Division. United States Strategic Bombing Survey (Pacific): Summary Report Covering Air Raid Protection and Allied Subjects in Japan (Report #10). Washington DC: U.S. Government Printing Office, 1946 (PDF Edition), 2.

38 Ness, Leland S., Rikugun: Guide to Japanese Ground Forces, 1937-1945; Tactical Organization of Imperial Japanese Army & Navy Ground Forces (Solihull: Helion & Company Ltd, 2014), 319-20; Zaloga, Steve. Defense of Japan, 1945. (Oxford: Osprey Publishing, 2010) (PDF Edition), 16-21.

39 Lardas, Mark. *Japan 1944-45: LeMay's B-29 Strategic Bombing Campaign*, (New York, Osprey Publishing, 2019)(PDF Edition), 27; Zaloga, Steve. *Defense of Japan, 1945*. Oxford: Osprey Publishing, 2010 (PDF Edition), 16-21.

40 Lardas, Mark. *Japan 1944-45: LeMay's B-29 Strategic Bombing Campaign*, (New York, Osprey Publishing, 2019)(PDF Edition), 27.

41 Lardas, Mark. *Japan 1944-45: LeMay's B-29 Strategic Bombing Campaign*, (New York, Osprey Publishing, 2019)(PDF Edition), 27, 29; Zaloga, Steve. *Defense of Japan, 1945*. Oxford: Osprey Publishing, 2010 (PDF Edition), 22-24.

42 Harris, Samuel R. Jr. (Robert A. Mann, Ed). *B-29s Over Japan 1944-45: A Group Commander's Diary*. (Jefferson, NC, McFarland & Co. Inc., 2011); Kohn, Richard H, *Strategic Air Warfare*. (Washington, DC, Office of Air Force History, United States Air Force, 1988), 53; Lardas, Mark. *Japan 1944-45: LeMay's B-29 Strategic Bombing Campaign*, (New York, Osprey Publishing, 2019)(PDF Edition), 27;

Zaloga, Steve. *Defense of Japan, 1945.* Oxford: Osprey Publishing, 2010 (PDF Edition), 54.

[43] Lardas, Mark. *Japan 1944-45: LeMay's B-29 Strategic Bombing Campaign*, (New York, Osprey Publishing, 2019)(PDF Edition), 27.

[44] Lardas, Mark. *Japan 1944-45: LeMay's B-29 Strategic Bombing Campaign*, (New York, Osprey Publishing, 2019)(PDF Edition), 27, 29.

45 Anderton, David A. *Aggressors, Volume 3: Interceptor vs. Heavy Bomber.* (Shrewsbury, UK: Airlife, 1991), 55, 50; Ikuhiko Hata, Izawa Yasuho, and Christopher F. Shores. X. *Japanese Army Fighter Aces, 1931-45.* (Mechanicsburg, PA: Stackpole Books, 2012) (PDF Edition), 147; Zaloga, Steve. *Defense of Japan, 1945.* Oxford: Osprey Publishing, 2010 (PDF Edition), 54.

[46] Lardas, Mark. *Japan 1944-45: LeMay's B-29 Strategic Bombing Campaign*, (New York, Osprey Publishing, 2019)(PDF Edition), 24; Nijboer, Donald. *B-29 Superfortress vs KI-44 "Tojo" (Duel).* (Oxford, England, Osprey Publishing, 2017 (E-book Version Bloomsbury Publishing)(Kindle Version), 3-5.

[47] XX Bomber Command, Twentieth Air Force. *B-29 Combat Crew Manual.* (India/China, 1944)(PDF Edition), X-6-X-10.

[48] Lardas, Mark. *Japan 1944-45: LeMay's B-29 Strategic Bombing Campaign*, (New York, Osprey Publishing, 2019)(PDF Edition), 29.

Before The Fire Blitz

[49] Nijboer, Donald. *B-29 Superfortress vs KI-44 "Tojo" (Duel).* (Oxford, England, Osprey Publishing, 2017 (E-book Version Bloomsbury Publishing)(Kindle Version), 34-5.

[50] Berger, Carl A. *B-29, The Superfortress.* (New York, Ballentine Books, 1970), 53-71.

[51] Coffey, Thomas M., *Iron Eagle: The Turbulent Life of General Curtis LeMay.* (New York: Avon Books, 1986), 110.

[52] Berger, Carl A. *B-29, The Superfortress.* (New York, Ballentine Books, 1970), 94-96; Craven, Wesley F. (ed). *United States Army Air Forces in World War II: Volume 5, The Pacific, Matterhorn to*

Nagasaki. (Washington, DC, US Printing Office, 1983)(PDF Edition), 144.

[53] Kohn, Richard H, *Strategic Air Warfare.* (AFD-100929-052)(Washington, DC, Office of Air Force History, United States Air Force, 1988)(PDF Edition), 67.

54 Kohn, Richard H, *Strategic Air Warfare.* (AFD-100929-052)(Washington, DC, Office of Air Force History, United States Air Force, 1988)(PDF Edition), 55 *ff*; Military Analysis Division, Army and Army Air Section. United States Strategic Bombing Survey: *The Strategic Air Operation of Very Heavy Bombardment in the War Against Japan (Twentieth Air Force) (Report 66).* (Washington, DC: U.S. Government Printing Office, 1946) (PDF Edition), 7-12.

[55] Snyder, Earl J. *General Leemy's Flying Circus: A Navigator's Story of the Twentieth Air Force in World War II.* (Madrid, Spain; P-47 Press, 2020)(Kindle Edition), 48.

[56] Tillitse, Lars. "When Bombs Rained On Us In Tokyo." *Saturday Evening Post* Magazine, 12 January 1946, 34.

[57] Harris, Samuel R. Jr. (Robert A. Mann, Ed). *B-29s Over Japan 1944-45: A Group Commander's Diary.* (Jefferson, NC, McFarland & Co. Inc., 2011)(Kindle Edition).

[58] Scott, James M. *Black Snow: Curtis LeMay, the Firebombing of Tokyo, and the Road to the Atomic Bomb.* (New York, WW Norton & Company, 2022)(Kindle Edition), 41.

[59] Scott, James M. *Black Snow: Curtis LeMay, the Firebombing of Tokyo, and the Road to the Atomic Bomb.* (New York, WW Norton & Company, 2022)(Kindle Edition), 44; Snyder, Earl J. *General Leemy's Flying Circus: A Navigator's Story of the Twentieth Air Force in World War II.* (Madrid, Spain; P-47 Press, 2020)(Kindle Edition), 64-65; Ugaki Matome, Donald M. Goldstein, Ed.. *Fading Victory: the Diary of Admiral Matatome Ugaki.* (Annapolis, MD, Naval Institute Press, 1991), 822.

[60] Harris, Samuel R. Jr. (Robert A. Mann, Ed). *B-29s Over Japan 1944-45: A Group Commander's Diary.* (Jefferson, NC, McFarland & Co. Inc., 2011)(Kindle Edition); Scott, James M. *Black Snow: Curtis LeMay, the Firebombing of Tokyo, and the Road to the Atomic Bomb.* (New York, WW Norton & Company, 2022)(Kindle Edition); Snyder, Earl J. *General Leemy's Flying Circus: A Navigator's Story*

of the Twentieth Air Force in World War II. (Madrid, Spain; P-47 Press, 2020)(Kindle Edition).

[61] XX Bomber Command, Twentieth Air Force. *B-29 Combat Crew Manual.* (India/China, 1944)(PDF Edition), XI-10, XIV-5.

[62] Scott, James M. *Black Snow: Curtis LeMay, the Firebombing of Tokyo, and the Road to the Atomic Bomb.* (New York, WW Norton & Company, 2022)(Kindle Edition), 59, 69.

[63] Harris, Samuel R. Jr. (Robert A. Mann, Ed). *B-29s Over Japan 1944-45: A Group Commander's Diary.* (Jefferson, NC, McFarland & Co. Inc., 2011)(Kindle Edition); Scott, James M. *Black Snow: Curtis LeMay, the Firebombing of Tokyo, and the Road to the Atomic Bomb.* (New York, WW Norton & Company, 2022)(Kindle Edition), 104-106.

64 Bradley, F. J. *No Strategic Targets Left.* Paducah, KY: Turner Publishing Company, 1999, 33; Craven, Wesley F. (ed). *United States Army Air Forces in World War II: Volume 5, The Pacific, Matterhorn to Nagasaki.* (Washington, DC, US Government Printing Office, 1983)(PDF Edition), 565.

65 Craven, Wesley F. (ed). *United States Army Air Forces in World War II: Volume 5, The Pacific, Matterhorn to Nagasaki.* (Washington, DC, US Printing Office, 1983)(PDF Edition), xviii; Dorr, Robert F. *Mission to Tokyo: The American Airmen Who Took the War to the Heart of Japan.* (Minneapolis: MBI Pub. Co., 2012), (Kindle Edition), 302; Kohn, Richard H, *Strategic Air Warfare. (AFD-100929-052)(*Washington, DC, Office of Air Force History, United States Air Force, 1988)(PDF Edition), 59.

[66] Harris, Samuel R. Jr.(Robert A. Mann, Ed). *B-29s Over Japan 1944-45: A Group Commander's Diary.* (Jefferson, NC, McFarland & Co. Inc., 2011)(Kindle Edition).

[67] Scott, James M. *Black Snow: Curtis LeMay, the Firebombing of Tokyo, and the Road to the Atomic Bomb.* (New York, WW Norton & Company, 2022)(Kindle Edition), 94, 104.

[68] Perera, Guido R. "Report To Committee Of Operations Analysts Economic Effects Of Successful Area Attacks On Six Japanese Cities." (Washington, DC, Committee of Operations Analysts, 1944)(PDF Edition); Scott, James M. *Black Snow: Curtis LeMay, the Firebombing of Tokyo, and the Road to the Atomic Bomb.* (New York, WW Norton & Company, 2022)(Kindle Edition), 101.

[69] Harris, Samuel R. Jr. (Robert A. Mann, Ed). *B-29s Over Japan 1944-45: A Group Commander's Diary.* (Jefferson, NC, McFarland & Co.

Inc., 2011)(Kindle Edition); USAAF "BG CE LeMay." Saipan, Mariana Islands, orders dated 20 January 1945. (PDF Edition).

70 Gorman, G. Scott. *Endgame in the Pacific: Complexity, Strategy and the B-29: A Fairchild Paper.* (Maxwell AFB, AL: Air University Press, February 2000) (PDF Edition), 10-11; LeMay, Curtis E. and Bill Yenne, *Superfortress: The Story of the B-29 and American Air Power.* (New York: McGraw-Hill, 1988) (PDF Edition), 103.

[71] Bond, Horatio, ed. Fire and the Air War; A Compilation of Expert Observations on Fires of the War Set by Incendiaries and the Atomic Bombs, Wartime Fire Fighting, and the Work of the Fire Protection Engineers Who Helped Plan and the Destruction of Enemy Cities and Industrial Plants. (Boston, MA: National Fire Protection Association International), 1946, 138, 141; Bradley, F. J. No Strategic Targets Left. (Paducah, KY: Turner Pub. Co., 1999), 33; Cahill, William L, "Imaging the Empire: The 3rd Photographic Reconnaissance Squadron in World War II." Air Force History, Spring 2012, 14; Mann, Robert A. The B-29 Superfortress: a Comprehensive Registry of the Planes and Their Missions. (Jefferson, NC: McFarland, 2009), 140-1; Tillman, Barrett. "When Fire Rained from the Sky." Aviation History, September 2016. http://www.historynet.com/ when-fire-rained-from-the-sky.htm, 34.

[72] Scott, James M. *Black Snow: Curtis LeMay, the Firebombing of Tokyo, and the Road to the Atomic Bomb.* (New York, WW Norton & Company, 2022)(Kindle Edition), 149-151; Snyder, Earl J. *General Leemy's Flying Circus: A Navigator's Story of the Twentieth Air Force in World War II.* (Madrid, Spain; P-47 Press, 2020)(Kindle Edition), 120.

[73] Snyder, Earl J. *General Leemy's Flying Circus: A Navigator's Story of the Twentieth Air Force in World War II.* (Madrid, Spain; P-47 Press, 2020)(Kindle Edition), 122.

[74] Craven, Wesley F. (ed). *United States Army Air Forces in World War II: Volume 5, The Pacific, Matterhorn to Nagasaki.* (Washington, DC, US Government Printing Office, 1983)(PDF Edition), 612, 613.

[75] Coffey, Thomas M. Iron Eagle: The Turbulent Life of General Curtis Lemay. New York: Crown Publishers, 1986), 138–45; Gorman, G Scott. Endgame in the Pacific: Complexity, Strategy and the B-29: A Fairchild Paper. Maxwell AFB, AL: Air University Press, February 2000 (PDF Edition), 13: Kohn, Richard H, Strategic Air Warfare.

(Washington, DC, Office of Air Force History, United States Air Force, 1988), 30.

[76] Scott, James M. *Black Snow: Curtis LeMay, the Firebombing of Tokyo, and the Road to the Atomic Bomb.* (New York, WW Norton & Company, 2022)(Kindle Edition), 183.

[77] Bradley, F. J. *No Strategic Targets Left.* Paducah, KY: Turner Publishing Company, 1999, 35; Coffey, Thomas M., *Iron Eagle: The Turbulent Life of General Curtis LeMay.* (New York: Crown Publishers), 1986, 139; LeMay, Curtis E. and Bill Yenne. *Superfortress: The Story of the B-29 and American Air Power.* (New York: McGraw-Hill, 1988) (PDF Edition), 122; Unknown, *9th Bomb Group War Diary Incendiary Raid List 4 February-17 June 1945.* (Unknown publisher, Unknown date)(PDF Edition), 1.

[78] Scott, James M. *Black Snow: Curtis LeMay, the Firebombing of Tokyo, and the Road to the Atomic Bomb.* (New York, WW Norton & Company, 2022)(Kindle Edition), 194-96.

79 Harris, Samuel R., *B-29s Over Japan, 1944-1945: A Group Commander's Diary*, ed. Robert A. Mann (Jefferson, NC: McFarland, 2011) (Kindle Edition); Kerr, E. Bartlett, *Flames Over Tokyo: The US Army Air Force's Incendiary Campaign against Japan 1944–45.* (New York: Donald I Fine, 1991), 158.

80 Kerr, E. Bartlett, *Flames Over Tokyo: The US Army Air Force's Incendiary Campaign against Japan 1944–45.* (New York: Donald I Fine), 1991, 158; LeMay, Curtis E. and Bill Yenne. *Superfortress: The Story of the B-29 and American Air Power.* New York: McGraw-Hill, 1988 (PDF Edition), 123; Rhodes, Richard, *The Making of the Atomic Bomb.* (New York: Simon & Schuster, 1986), 597.

81 Kerr, E. Bartlett, *Flames Over Tokyo: The US Army Air Force's Incendiary Campaign against Japan 1944–45.* (New York: Donald I Fine), 1991, 157, 161.

82 Cadin, Martin, *A Torch to the Enemy.* (New York: Ballentine Books, 1960), 100; Dorr, Robert F. *Mission to Tokyo: the American Airmen Who Took the War to the Heart of Japan.* (Minneapolis: MBI Pub. Co., 2012) (Kindle Edition); Kerr, E. Bartlett, *Flames Over*

Tokyo: The US Army Air Force's Incendiary Campaign against Japan 1944–45 (New York: Donald I Fine), 1991, 161.

[83] Scott, James M. *Black Snow: Curtis LeMay, the Firebombing of Tokyo, and the Road to the Atomic Bomb.* (New York, WW Norton & Company, 2022)(Kindle Edition), 201, 217.

[84] Scott, James M. *Black Snow: Curtis LeMay, the Firebombing of Tokyo, and the Road to the Atomic Bomb.* (New York, WW Norton & Company, 2022)(Kindle Edition), 196.

The Fire Blitz Begins

85 Coffey, Thomas M., *Iron Eagle: The Turbulent Life of General Curtis Lemay.* (New York: Crown Publishers, 1986), 148; Kohn, Richard H, *Strategic Air Warfare.* (Washington, DC, Office of Air Force History, United States Air Force, 1988), 54; Harris, Samuel R. Jr. (Robert A. Mann, Ed). *B-29s Over Japan 1944-45: A Group Commander's Diary.* (Jefferson, NC, McFarland & Co. Inc., 2011); LeMay, Curtis E. and Bill Yenne. *Superfortress: The Story of the B-29 and American Air Power.* New York: (McGraw-Hill, 1988) (PDF Edition), 123; Kerr, E. Bartlett, *Flames Over Tokyo: The US Army Air Force's Incendiary Campaign against Japan 1944–45.* (New York: Donald I Fine), 1991, 166.

86 Associated Press. "Deadly WWII U.S. Firebombing Raids on Japanese Cities Largely Ignored." *The Japan Times.* Japan Times Limited, March 10, 2015. Accessed October 5, 2017. https://www.japantimes.co.jp/news/ 2015/03/10/ national/deadly—wwii-u-s—firebombing-raids-on-japanese-cities—largely—ignored/#.WdYoqPKJA; Edoin Hoito, *The Night Tokyo Burned: The Incendiary Campaign against Japan, March-August 1945.* (New York: St. Martin's Press, 1987), 63; Kerr, E. Bartlett, *Flames Over Tokyo: The US Army Air Force's Incendiary Campaign against Japan 1944–45.* (New York: Donald I Fine, 1991), 152, 153.

87 Snyder, Earl J. *General Leemy's Flying Circus: A Navigator's Story of the Twentieth Air Force in World War II.* (Madrid, Spain; P-47 Press, 2020)(Kindle Edition), 125; Kerr, E. Bartlett, *Flames Over Tokyo: The US Army Air Force's Incendiary Campaign against Japan 1944–45.* (New York: Donald I Fine, 1991), 167; Tillman, Barrett, "Cities Reduced to Ashes." (Washington, DC,

American Heritage Magazine Volume 68, Number 5, August 2023), 6.

[88] Scott, James M. *Black Snow: Curtis LeMay, the Firebombing of Tokyo, and the Road to the Atomic Bomb.* (New York, WW Norton & Company, 2022)(Kindle Edition), 235.

[89] Scott, James M. *Black Snow: Curtis LeMay, the Firebombing of Tokyo, and the Road to the Atomic Bomb.* (New York, WW Norton & Company, 2022)(Kindle Edition), 231.

[90] Tillman, Barrett, "Cities Reduced to Ashes." (Washington, DC, *American Heritage Magazine* Volume 68, Number 5, August 2023), 6.

[91] Scott, James M. *Black Snow: Curtis LeMay, the Firebombing of Tokyo, and the Road to the Atomic Bomb.* (New York, WW Norton & Company, 2022)(Kindle Edition), 514-15.

92 Cadin, Martin, *A Torch to the Enemy.* (New York: Ballentine Books, 1960), 122; Kerr, E. Bartlett, *Flames Over Tokyo: The US Army Air Force's Incendiary Campaign against Japan 1944–45.* (New York: Donald I Fine, 1991), 171, 178.

93 Craven, Wesley F. (ed). *United States Army Air Forces in World War II: Volume 5, The Pacific, Matterhorn to Nagasaki.* (Washington, DC, US Government Printing Office, 1983)(PDF Edition), 614; Rhodes, Richard, *The Making of the Atomic Bomb.* (New York: Simon & Schuster, 1986), 597; Scott, James M. *Black Snow: Curtis LeMay, the Firebombing of Tokyo, and the Road to the Atomic Bomb.* (New York, WW Norton & Company, 2022)(Kindle Edition), 242; Tillman, Barrett. "When Fire Rained from the Sky." *Aviation History*, September 2016. http://www.historynet.com /when-fire-rained-from-the-sky.htm, 31.

[94] Craven, Wesley F. (ed). *United States Army Air Forces in World War II: Volume 5, The Pacific, Matterhorn to Nagasaki.* (Washington, DC, US Government Printing Office, 1983)(PDF Edition), 615.

95 Edoin, Hoito, *The Night Tokyo Burned: The Incendiary Campaign against Japan, March-August 1945.* (New York: St. Martin's Press, 1987), 63–64.

[96] Craven, Wesley F. (ed). *United States Army Air Forces in World War II: Volume 5, The Pacific, Matterhorn to Nagasaki.* (Washington, DC, US Government Printing Office, 1983)(PDF Edition), 615.

97 Cadin, Martin, *A Torch to the Enemy.* (New York: Ballentine Books, 1960), 108; Cook, Haruko Taya and Theodore F.

Cook. *Japan at War: An Oral History.* (New York: New Press, 1993), 345; Edoin Hoito, *The Night Tokyo Burned: The Incendiary Campaign against Japan, March-August 1945.* (New York: St. Martin's Press, 1987), 21.

98 Craven, Wesley F. (ed). *United States Army Air Forces in World War II: Volume 5, The Pacific, Matterhorn to Nagasaki.* (Washington, DC, US Government Printing Office, 1983)(PDF Edition), 614; Scott, James M. *Black Snow: Curtis LeMay, the Firebombing of Tokyo, and the Road to the Atomic Bomb.* (New York, WW Norton & Company, 2022)(Kindle Edition), 244; Rhodes, Richard, *The Making of the Atomic Bomb.* (New York: Simon & Schuster, 1986), 597; Tillman, Barrett, "Cities Reduced to Ashes." (Washington, DC, *American Heritage Magazine* Volume 68, Number 5, August 2023), 5.

99 Associated Press. "Deadly WWII U.S. Firebombing Raids on Japanese Cities Largely Ignored." *The Japan Times.* (Japan Times Limited, March 10, 2015). Last modified March 10, 2015. Accessed October 5, 2017. https://www.japantimes.co.jp/news/2015/03/10/national/deadly—wwii-u-s—firebombing-raids-on-japanese-cities—largely—ignored/#.WdYoqPKJA.

100 Associated Press. "Deadly WWII U.S. Firebombing Raids on Japanese Cities Largely Ignored." *The Japan Times.* Japan Times Limited, March 10, 2015. Last modified March 10, 2015. Accessed October 5, 2017. https://www.japantimes.co.jp/news/2015/03/10/national/deadly—wwii-u-s—firebombing-raids-on-japanese-cities—largely—ignored/#.WdYoqPKJA.

101 Bradley, F. J. *No Strategic Targets Left.* (Paducah, KY: Turner Publishing Company, 1999), 34, 43.

102 Bradley, F. J. *No Strategic Targets Left.* (Paducah, KY: Turner Publishing Company, 1999), 34, 43; Cadin, Martin, *A Torch to the Enemy.* (New York: Ballentine Books, 1960), 111.

103 Bond, Horatio, ed. Fire and the Air War; A Compilation of Expert Observations on Fires of the War Set by Incendiaries and the Atomic Bombs, Wartime Fire Fighting, and the Work of the Fire Protection Engineers Who Helped Plan and the Destruction of Enemy Cities and Industrial Plants. (Boston, MA: National Fire Protection Association International, 1946), 138; Craven, Wesley F. (ed). United States Army Air Forces in World War II: Volume 5, The Pacific, Matterhorn to Nagasaki. (Washington, DC, US Government Printing Office, 1983)(PDF Edition), 617; Guillain, Robert. I Saw Tokyo

Burning: An Eyewitness Narrative from Pearl Harbor to Hiroshima. (William Byron, trans.)(New York, Ballentine Books, 1981), 185.

[104] Craven, Wesley F. (ed). *United States Army Air Forces in World War II: Volume 5, The Pacific, Matterhorn to Nagasaki.* (Washington, DC, US Government Printing Office, 1983)(PDF Edition), 616.

105 Tillman, Barrett. "When Fire Rained from the Sky." *Aviation History*, September 2016. http://www.historynet.com /when-fire-rained-from-the-sky.htm, 33.

106 Bradley, F. J. *No Strategic Targets Left.* (Paducah, KY: Turner Publishing Company, 1999), 34; Kerr, E. Bartlett, *Flames Over Tokyo: The US Army Air Force's Incendiary Campaign against Japan 1944–45.* (New York: Donald I Fine, 1991), 181.

107 Craven, Wesley F. (ed). *United States Army Air Forces in World War II: Volume 5, The Pacific, Matterhorn to Nagasaki.* (Washington, DC, US Government Printing Office, 1983)(PDF Edition), 616; Ikuhiko Hata, Izawa Yasuho, and Christopher F. Shores. X. *Japanese Army Fighter Aces, 1931-45.* (Mechanicsburg, PA: Stackpole Books, 2012), 147.

108 Associated Press. "Deadly WWII U.S. Firebombing Raids on Japanese Cities Largely Ignored." *The Japan Times*. Japan Times Limited, March 10, 2015. Last modified March 10, 2015. Accessed October 5, 2017. https://www.japantimes.co.jp/news/ 2015/03/10/national/deadly—wwii-u-s—firebombing-raids-on-japanese-cities—largely—ignored/#.WdYoqPKJA.

109 Bond, Horatio, ed. Fire and the Air War; a Compilation of Expert Observations on Fires of the War Set by Incendiaries and the Atomic Bombs, Wartime Fire Fighting, and the Work of the Fire Protection Engineers Who Helped Plan and the Destruction of Enemy Cities and Industrial Plants. (Boston, MA: National Fire Protection Association

International, 1946), 7; Cadin, Martin, A Torch to the Enemy. (New York: Ballentine Books, 1960), 118.

110 Tillman, Barrett. "When Fire Rained from the Sky." *Aviation History*, September 2016. http://www.historynet.com /when-fire-rained-from-the-sky.htm, 31.

111. Edoin Hoito, *The Night Tokyo Burned: The Incendiary Campaign against Japan, March-August 1945.* (New York: St. Martin's Press, 1987), 65.

112. Kerr, E. Bartlett, *Flames Over Tokyo: The US Army Air Force's Incendiary Campaign against Japan 1944–45.* (New York: Donald I Fine, 1991), 182.

113 Edoin Hoito, *The Night Tokyo Burned: The Incendiary Campaign against Japan, March-August 1945.* (New York: St. Martin's Press, 1987), 65.

114 Cook, Haruko Taya and Theodore F. Cook. *Japan at War: An Oral History.* (New York: New Press, 1993), 352; Edoin Hoito, *The Night Tokyo Burned: The Incendiary Campaign against Japan, March-August 1945.* (New York: St. Martin's Press, 1987), 82–84; Kerr, E. Bartlett, *Flames Over Tokyo: The US Army Air Force's Incendiary Campaign against Japan 1944–45.* (New York: Donald I Fine, 1991), 201.

115 Tillman, Barrett. "When Fire Rained from the Sky." *Aviation History*, September 2016. http://www.historynet.com /when-fire-rained-from-the-sky.htm, 33.

116 Bradley, F. J. *No Strategic Targets Left.* (Paducah, KY: Turner Publishing Company, 1999), 34, 43.

[117] Tillman, Barrett, "Cities Reduced to Ashes." (Washington, DC, *American Heritage Magazine* Volume 68, Number 5, August 2023), 12.

[118] Tillman, Barrett, "Cities Reduced to Ashes." (Washington, DC, *American Heritage Magazine* Volume 68, Number 5, August 2023), 11.

119 Cook, Haruko Taya and Theodore F. Cook. *Japan at War: An Oral History.* (New York: New Press, 1993), 348-9; Tillman, Barrett. "When Fire Rained from the Sky." *Aviation History*, September

2016. http://www.historynet.com /when-fire-rained-from-the-sky.htm, 33.

120 Tillman, Barrett. "When Fire Rained from the Sky." *Aviation History*, September 2016. http://www.historynet.com /when-fire-rained-from-the-sky.htm, 34.

[121] Associated Press. "Deadly WWII U.S. Firebombing Raids on Japanese Cities Largely Ignored." *The Japan Times*. Japan Times Limited, March 10, 2015. Last modified March 10, 2015. Accessed October 5, 2017. https://www.japantimes.co.jp/news/2015/03/10/national/deadly—wwii-u-s—firebombing-raids-on-japanese-cities—largely—ignored/#.WdYoqPKJA.

[122] Associated Press. "Deadly WWII U.S. Firebombing Raids on Japanese Cities Largely Ignored." *The Japan Times*. Japan Times Limited, March 10, 2015. Last modified March 10, 2015. Accessed October 5, 2017. https://www.japantimes.co.jp/news/2015/03/10/national/deadly—wwii-u-s—firebombing-raids-on-japanese-cities—largely—ignored/#.WdYoqPKJA.

[123] Tillman, Barrett, "Cities Reduced to Ashes." (Washington, DC, *American Heritage Magazine* Volume 68, Number 5, August 2023), 13.

124 Bond, Horatio, ed. Fire and the Air War; A Compilation of Expert Observations on Fires of the War Set by Incendiaries and the Atomic Bombs, Wartime Fire Fighting, and the Work of the Fire Protection Engineers Who Helped Plan and the Destruction of Enemy Cities and Industrial Plants. (Boston, MA: National Fire Protection Association International, 1946), 165; Edoin Hoito, The Night Tokyo Burned: The Incendiary Campaign against Japan, March-August 1945. (New York: St. Martin's Press, 1987), 106; Tillman, Barrett. "When Fire Rained from the Sky." Aviation History, September 2016. http://www.historynet.com /when-fire-rained-from-the-sky.htm, 35.

[125] Scott, James M. *Black Snow: Curtis LeMay, the Firebombing of Tokyo, and the Road to the Atomic Bomb.* (New York, WW Norton & Company, 2022)(Kindle Edition), 277.

[126] Craven, Wesley F. (ed). *United States Army Air Forces in World War II: Volume 5, The Pacific, Matterhorn to Nagasaki.* (Washington, DC, US Government Printing Office, 1983)(PDF Edition), 616; Scott, James M. *Black Snow: Curtis LeMay, the*

Firebombing of Tokyo, and the Road to the Atomic Bomb. (New York, WW Norton & Company, 2022)(Kindle Edition), 289, 305.

127 Tillman, Barrett. "When Fire Rained from the Sky." *Aviation History*, September 2016. http://www.historynet.com /when-fire-rained-from-the-sky.htm, 33. Tillman, Barrett, "Cities Reduced to Ashes." (Washington, DC, *American Heritage Magazine* Volume 68, Number 5, August 2023), 9-10.

128 *NewspaperArchive.com*, accessed March 27, 2019.

129 Craven, Wesley F. (ed). *United States Army Air Forces in World War II: Volume 5, The Pacific, Matterhorn to Nagasaki.* (Washington, DC, US Government Printing Office, 1983)(PDF Edition), 617; Kerr, E. Bartlett. *Flames over Tokyo: the U.S. Army Air Force's Incendiary Campaign against Japan, 1944-1945* (New York: D.I. Fine, 1991), 210-11. 214; Tillman, Barrett, "Cities Reduced to Ashes." (Washington, DC, *American Heritage Magazine* Volume 68, Number 5, August 2023), 15.

[130] Scott, James M. *Black Snow: Curtis LeMay, the Firebombing of Tokyo, and the Road to the Atomic Bomb.* (New York, WW Norton & Company, 2022)(Kindle Edition), 302.

[131] Daniels, Gordon, "Before Hiroshima: The Bombing of Japan, 1944-45." *History Today*, January 1982, 18; Ugaki Matome, Donald M. Goldstein, Ed.. *Fading Victory: the Diary of Admiral Matatome Ugaki.*(Annapolis, MD, Naval Institute Press, 1991), 548-9.

[132] Craven, Wesley F. (ed). *United States Army Air Forces in World War II: Volume 5, The Pacific, Matterhorn to Nagasaki.* (Washington, DC, US Government Printing Office, 1983)(PDF Edition), 617.

The Fire Blitz Goes On

133 LeMay, Curtis E. and Bill Yenne. *Superfortress: The Story of the B-29 and American Air Power.* New York: McGraw-Hill, 1988 (PDF Edition), 123.

134 Harris, Samuel R., *B-29s Over Japan, 1944-1945: A Group Commander's Diary*, ed. Robert A. Mann (Jefferson, NC: McFarland, 2011) (Kindle Edition).

[135] Craven, Wesley F. (ed). *United States Army Air Forces in World War II: Volume 5, The Pacific, Matterhorn to Nagasaki.*

(Washington, DC, US Government Printing Office, 1983)(PDF Edition), 618.

[136] Craven, Wesley F. (ed). *United States Army Air Forces in World War II: Volume 5, The Pacific, Matterhorn to Nagasaki.* (Washington, DC, US Government Printing Office, 1983)(PDF Edition), 618; *Impact Magazine*, Volume III, Number 4 (June 1945), 30.

137 Bradley, F. J. *No Strategic Targets Left.* (Paducah, KY: Turner Publishing Company, 1999), 35, 43, 46; Craven, Wesley F. (ed). *United States Army Air Forces in World War II: Volume 5, The Pacific, Matterhorn to Nagasaki.* (Washington, DC, US Government Printing Office, 1983)(PDF Edition), 619; LeMay, Curtis E. and Bill Yenne. *Superfortress: The Story of the B-29 and American Air Power.*)New York: McGraw-Hill, 1988), (PDF Edition), 106; Military Analysis Division, Army and Army Air Section. *United States Strategic Bombing Survey: The Effects of Incendiary Bomb Attacks on Japan (Report 90).* (Washington, DC: U.S. Government Printing Office, 1946) (PDF Edition), 7; Stevens, Mark. "XXI Bomber Command Mission Summaries." *20th Air Force Mission Summaries.* Twentieth Air Force Association, 2004. Last modified 2004. Accessed October 5, 2017 http://www.20thaf.org/missions /missions_index.htm; Tillman, Barrett. "When Fire Rained from the Sky." *Aviation History*, September 2016. http://www.historynet.com /when-fire-rained-from-the-sky.htm, 35; Werrell, Kenneth P. *Blankets of Fire: U.S. Bombers over Japan during World War II.* (Washington, DC: Smithsonian institution Press, 1996), 164.

138 Craven, Wesley F. (ed). *United States Army Air Forces in World War II: Volume 5, The Pacific, Matterhorn to Nagasaki.* (Washington, DC, US Government Printing Office, 1983)(PDF Edition), 619; LeMay, Curtis E. and Bill Yenne. *Superfortress: The Story of the B-29 and American Air Power.* (New York: McGraw-Hill, 1988) (PDF Edition), 106. Stevens, Mark. "XXI Bomber Command Mission Summaries." *20th Air Force Mission Summaries.* Twentieth Air Force Association, 2004. Last modified 2004. Accessed October 5, 2017. http://www.20thaf.org/missions /missions_index.htm; Werrell, Kenneth P. *Blankets of Fire: U.S. Bombers over Japan during World War II.* (Washington, DC: Smithsonian Institution Press, 1996). 163-164.

[139] Craven, Wesley F. (ed). *United States Army Air Forces in World War II: Volume 5, The Pacific, Matterhorn to Nagasaki.*

(Washington, DC, US Government Printing Office, 1983)(PDF Edition), 618.

[140] *Impact Magazine*, Volume III, Number 4 (June 1945), 26.

[141] Craven, Wesley F. (ed). *United States Army Air Forces in World War II: Volume 5, The Pacific, Matterhorn to Nagasaki.* (Washington, DC, US Government Printing Office, 1983)(PDF Edition), 619-620.

[142] Harris, Samuel R. Jr. (Robert A. Mann, Ed). *B-29s Over Japan 1944-45: A Group Commander's Diary.* (Jefferson, NC, McFarland & Co. Inc., 2011)(Kindle Edition).

[143] Craven, Wesley F. (ed). *United States Army Air Forces in World War II: Volume 5, The Pacific, Matterhorn to Nagasaki.* (Washington, DC, US Government Printing Office, 1983)(PDF Edition), 620.

144 Snyder, Earl J. *General Leemy's Flying Circus: A Navigator's Story of the Twentieth Air Force in World War II.* (Madrid, Spain; P-47 Press, 2020)(Kindle Edition), 127-129; *Air Intelligence Report*, 3rd ed., vol. 1 (APO 23: Twentieth Air Force, 1945) (PDF Edition), 7; Bradley, F. J. *No Strategic Targets Left.* Paducah, KY: Turner Publishing Company, 1999, 35.

145 LeMay, Curtis E. and Bill Yenne. *Superfortress: The Story of the B-29 and American Air Power.* New York: McGraw-Hill, 1988, (PDF Edition) 106.

146 Bond, Horatio, ed. Fire and the Air War; A Compilation of Expert Observations on Fires of the War Set by Incendiaries and the Atomic Bombs, Wartime Fire Fighting, and the Work of the Fire Protection Engineers Who Helped Plan and the Destruction of Enemy Cities and Industrial Plants. (Boston, MA: National Fire Protection Association International, 1946), 160; Bradley, F. J. No Strategic Targets Left. (Paducah, KY: Turner Publishing Company, 1999), 35; Craven, Wesley F. (ed). United States Army Air Forces in World War II: Volume 5, The Pacific, Matterhorn to Nagasaki. (Washington, DC, US Government Printing Office, 1983)(PDF Edition), 619-621; Impact Magazine, Volume III, Number 4 (June 1945), 26; Tillman, Barrett. "When Fire Rained from the Sky." Aviation History,

September 2016. http://www.historynet.com /when-fire-rained-from-the-sky.htm, 35.

[147] Scott, James M. *Black Snow: Curtis LeMay, the Firebombing of Tokyo, and the Road to the Atomic Bomb.* (New York, WW Norton & Company, 2022)(Kindle Edition), 298.

[148] Scott, James M. *Black Snow: Curtis LeMay, the Firebombing of Tokyo, and the Road to the Atomic Bomb.* (New York, WW Norton & Company, 2022)(Kindle Edition), 299-300.

[149] Craven, Wesley F. (ed). *United States Army Air Forces in World War II: Volume 5, The Pacific, Matterhorn to Nagasaki.* (Washington, DC, US Government Printing Office, 1983)(PDF Edition), 617.

[150] Snyder, Earl J. *General Leemy's Flying Circus: A Navigator's Story of the Twentieth Air Force in World War II.* (Madrid, Spain; P-47 Press, 2020)(Kindle Edition), 151.

151 Bradley, F. J. *No Strategic Targets Left.* (Paducah, KY: Turner Publishing Company, 1999), 35; Craven, Wesley F. (ed). *United States Army Air Forces in World War II: Volume 5, The Pacific, Matterhorn to Nagasaki.* (Washington, DC, US Government Printing Office, 1983)(PDF Edition), 621-622; *Impact Magazine*, Volume III, Number 4 (June 1945), 28-29; Stevens, Mark. "XXI Bomber Command Mission Summaries." *20th Air Force Mission Summaries.* Twentieth Air Force Association, 2004. Last modified 2004. Accessed October 5, 2017. http://www.20thaf.org/missions/ missions_index.htm; Takai Koji, and Henry Sakaida, *B-29 Hunters of the JAAF.* (Oxford, England, Osprey Publishing, 2001)(PDF Edition), 91.

[152] Takai Koji, and Henry Sakaida, *B-29 Hunters of the JAAF.* (Oxford, England, Osprey Publishing, 2001)(PDF Edition), 92.

[153] Takai Koji, and Henry Sakaida, *B-29 Hunters of the JAAF.* (Oxford, England, Osprey Publishing, 2001)(PDF Edition), 94.

[154] Snyder, Earl J. *General Leemy's Flying Circus: A Navigator's Story of the Twentieth Air Force in World War II.* (Madrid, Spain; P-47 Press, 2020)(Kindle Edition), 162-166.

155 Bradley, F. J. *No Strategic Targets Left.* (Paducah, KY: Turner Publishing Company, 1999), 36; Kerr, E. Bartlett. *Flames over Tokyo: the U.S. Army Air Force's Incendiary Campaign against Japan, 1944-1945* (New York: D.I. Fine, 1991), 218-19; Scott, James M. *Black Snow: Curtis LeMay, the Firebombing of Tokyo, and the*

Road to the Atomic Bomb. (New York, WW Norton & Company, 2022)(Kindle Edition), 311.

[156] LeMay, Curtis E. and Bill Yenne. *Superfortress: The Story of the B-29 and American Air Power.* (New York: McGraw-Hill, 1988)(PDF Edition), 125; Scott, James M. *Black Snow: Curtis LeMay, the Firebombing of Tokyo, and the Road to the Atomic Bomb.* (New York, WW Norton & Company, 2022)(Kindle Edition), 301.

[157] Snyder, Earl J. *General Leemy's Flying Circus: A Navigator's Story of the Twentieth Air Force in World War II.* (Madrid, Spain; P-47 Press, 2020)(Kindle Edition), 164-168.

[158] Scott, James M. *Black Snow: Curtis LeMay, the Firebombing of Tokyo, and the Road to the Atomic Bomb.* (New York, WW Norton & Company, 2022)(Kindle Edition), 302.

[159] Scott, James M. *Black Snow: Curtis LeMay, the Firebombing of Tokyo, and the Road to the Atomic Bomb.* (New York, WW Norton & Company, 2022)(Kindle Edition), 340-1.

160 LeMay, Curtis E. and Bill Yenne. *Superfortress: The Story of the B-29 and American Air Power.* (New York: McGraw-Hill, 1988) (PDF Edition) 106.; "WWII Japanese Leaflets," *WallBuilders*, last modified January 18, 2017, accessed March 26, 2019, https://wallbuilders.com/wwii-japanese-leaflets/#.

161 Military Analysis Division, Army and Army Air Section. United States Strategic Bombing Survey: The Strategic Air Operation of Very Heavy Bombardment in the War Against Japan (Twentieth Air Force) (Report 66). Washington, DC: (U.S. Government Printing Office, 1946) (PDF Edition), 15

162 Bradley, F. J. *No Strategic Targets Left.* (Paducah, KY: Turner Publishing Company, 1999), 36; Rhodes, Richard, *The Making of the Atomic Bomb.* (New York: Simon & Schuster, 1986), 600; Tillman, Barrett. "When Fire Rained from the Sky." *Aviation History,*

September 2016. http://www.historynet.com /when-fire-rained-from-the-sky.htm, 35.

[163] Ugaki Matome, Donald M. Goldstein, Ed.. *Fading Victory: the Diary of Admiral Matatome Ugaki.*(Annapolis, MD, Naval Institute Press, 1991), 562-3.

[164] *Impact Magazine*, Volume III, Number 4 (June 1945), 34.

[165] Scott, James M. *Black Snow: Curtis LeMay, the Firebombing of Tokyo, and the Road to the Atomic Bomb.* (New York, WW Norton & Company, 2022)(Kindle Edition), 318-19.

[166] Craven, Wesley F. (ed). *United States Army Air Forces in World War II: Volume 5, The Pacific, Matterhorn to Nagasaki.* (Washington, DC, US Government Printing Office, 1983)(PDF Edition), 636; Scott, James M. *Black Snow: Curtis LeMay, the Firebombing of Tokyo, and the Road to the Atomic Bomb.* (New York, WW Norton & Company, 2022)(Kindle Edition), 321; Unknown, *9th Bomb Group War Diary Incendiary Raid List 4 February-17 June 1945.* (Unknown publisher, Unknown date)(PDF Edition), 2.

[167] Ugaki Matome, Donald M. Goldstein, Ed.. *Fading Victory: the Diary of Admiral Matatome Ugaki.*(Annapolis, MD, Naval Institute Press, 1991), 584.

[168] Craven, Wesley F. (ed). *United States Army Air Forces in World War II: Volume 5, The Pacific, Matterhorn to Nagasaki.* (Washington, DC, US Government Printing Office, 1983)(PDF Edition), 636-637; Scott, James M. *Black Snow: Curtis LeMay, the Firebombing of Tokyo, and the Road to the Atomic Bomb.* (New York, WW Norton & Company, 2022)(Kindle Edition), 321.

[169] Craven, Wesley F. (ed). *United States Army Air Forces in World War II: Volume 5, The Pacific, Matterhorn to Nagasaki.* (Washington, DC, US Government Printing Office, 1983)(PDF Edition), 637-638; Ugaki Matome, Donald M. Goldstein, Ed.. *Fading Victory: the Diary of Admiral Matatome Ugaki.*(Annapolis, MD, Naval Institute Press, 1991), 613.

[170] Bradley, F. J. *No Strategic Targets Left.* (Paducah, KY: Turner Publishing Company, 1999), 37; Craven, Wesley F. (ed). *United States Army Air Forces in World War II: Volume 5, The Pacific, Matterhorn to Nagasaki.* (Washington, DC, US Government Printing Office, 1983)(PDF Edition), 638; Scott, James M. *Black Snow: Curtis LeMay, the Firebombing of Tokyo, and the Road to the Atomic Bomb.* (New York, WW Norton & Company, 2022)(Kindle Edition),

321-2; Ugaki Matome, Donald M. Goldstein, Ed.. *Fading Victory: the Diary of Admiral Matatome Ugaki.*(Annapolis, MD, Naval Institute Press, 1991), 616.

[171] Scott, James M. *Black Snow: Curtis LeMay, the Firebombing of Tokyo, and the Road to the Atomic Bomb.* (New York, WW Norton & Company, 2022)(Kindle Edition), 323.

[172] Craven, Wesley F. (ed). *United States Army Air Forces in World War II: Volume 5, The Pacific, Matterhorn to Nagasaki.* (Washington, DC, US Government Printing Office, 1983)(PDF Edition), 638-639; Ugaki Matome, Donald M. Goldstein, Ed.. *Fading Victory: the Diary of Admiral Matatome Ugaki.*(Annapolis, MD, Naval Institute Press, 1991), 620; Unknown, *9th Bomb Group War Diary Incendiary Raid List 4 February-17 June 1945.* (Unknown publisher, Unknown date)(PDF Edition), 2.

[173] Scott, James M. *Black Snow: Curtis LeMay, the Firebombing of Tokyo, and the Road to the Atomic Bomb.* (New York, WW Norton & Company, 2022)(Kindle Edition), 335.

[174] Selden, Mark. "Bombs Bursting in Air: State and Citizen Responses to the US Firebombing and Atomic Bombing of Japan." *Asia-Pacific Journal*, Vol. 12, Issue 3, No. 4, 14 January 2014, 4.

[175] Ugaki Matome, Donald M. Goldstein, Ed.. *Fading Victory: the Diary of Admiral Matatome Ugaki.*(Annapolis, MD, Naval Institute Press, 1991), 576.

[176] Craven, Wesley F. (ed). *United States Army Air Forces in World War II: Volume 5, The Pacific, Matterhorn to Nagasaki.* (Washington, DC, US Government Printing Office, 1983)(PDF Edition), 639-640; Unknown, *9th Bomb Group War Diary Incendiary Raid List 4 February-17 June 1945.* (Unknown publisher, Unknown date)(PDF Edition), 2.

[177] Scott, James M. *Black Snow: Curtis LeMay, the Firebombing of Tokyo, and the Road to the Atomic Bomb.* (New York, WW Norton & Company, 2022)(Kindle Edition), 323.

[178] Craven, Wesley F. (ed). *United States Army Air Forces in World War II: Volume 5, The Pacific, Matterhorn to Nagasaki.* (Washington, DC, US Government Printing Office, 1983)(PDF Edition), 640-641; Ugaki Matome, Donald M. Goldstein, Ed.. *Fading Victory: the Diary of Admiral Matatome Ugaki.*(Annapolis, MD, Naval Institute Press, 1991), 624; Unknown, *9th Bomb Group War*

Diary Incendiary Raid List 4 February-17 June 1945. (Unknown publisher, Unknown date)(PDF Edition), 2.

[179] Tillman, Barrett. "506th Fighter Group - Mustangs of Iwo." *Airpower Magazine*, Vol. 7, #1, 1977. https://www.506thfightergroup.org/ mustangsofiwo.asp. Accessed December 2023.

[180] Craven, Wesley F. (ed). *United States Army Air Forces in World War II: Volume 5, The Pacific, Matterhorn to Nagasaki.* (Washington, DC, US Government Printing Office, 1983)(PDF Edition), 642; Unknown, *9th Bomb Group War Diary Incendiary Raid List 4 February-17 June 1945.* (Unknown publisher, Unknown date)(PDF Edition), 2.

[181] Craven, Wesley F. (ed). *United States Army Air Forces in World War II: Volume 5, The Pacific, Matterhorn to Nagasaki.* (Washington, DC, US Government Printing Office, 1983)(PDF Edition), 642-43; Ugaki Matome, Donald M. Goldstein, Ed.. *Fading Victory: the Diary of Admiral Matatome Ugaki.*(Annapolis, MD, Naval Institute Press, 1991), 633; Unknown, *9th Bomb Group War Diary Incendiary Raid List 4 February-17 June 1945.* (Unknown publisher, Unknown date)(PDF Edition), 2.

[182] Craven, Wesley F. (ed). *United States Army Air Forces in World War II: Volume 5, The Pacific, Matterhorn to Nagasaki.* (Washington, DC, US Government Printing Office, 1983)(PDF Edition), 643-44.

183 Bradley, F. J. *No Strategic Targets Left.* (Paducah, KY: Turner Publishing Company, 1999), 38.

The Little Fire Blitz

[184] Scott, James M. *Black Snow: Curtis LeMay, the Firebombing of Tokyo, and the Road to the Atomic Bomb.* (New York, WW Norton & Company, 2022)(Kindle Edition), 324-5.

[185] Ugaki Matome, Donald M. Goldstein, Ed.. *Fading Victory: the Diary of Admiral Matatome Ugaki.*(Annapolis, MD, Naval Institute Press, 1991), 634.

186 Bradley, F. J. *No Strategic Targets Left.* (Paducah, KY: Turner Publishing Company, 1999), 36, 37-38; Cadin, Martin, *A Torch to the Enemy.* (New York: Ballentine Books, 1960), 158; Craven, Wesley F. (ed). *United States Army Air Forces in World War II:*

Volume 5, The Pacific, Matterhorn to Nagasaki. (Washington, DC, US Government Printing Office, 1983)(PDF Edition), 653-55.

[187] Craven, Wesley F. (ed). *United States Army Air Forces in World War II: Volume 5, The Pacific, Matterhorn to Nagasaki.* (Washington, DC, US Government Printing Office, 1983)(PDF Edition), 658.

Vindication, Vengeance, and Marketing

[188] Whitman, John W.. "Japan's Fatally Flawed Air Forces in World War II." https://www.historynet.com/japans-fatally-flawed-air-forces-in-world-war-ii-2/ Accessed 13 Sep 23.

189 Military Analysis Division, Army and Army Air Section. *United States Strategic Bombing Survey: The Effects of Strategic Bombing on Japan's War Economy (Report 53).* (Washington, DC: U.S. Government Printing Office, 1947) (PDF Edition), 63.

190 Military Analysis Division, Army and Army Air Section. *United States Strategic Bombing Survey: The Effects of Strategic Bombing on Japan's War Economy (Report 53).* (Washington, DC: U.S. Government Printing Office, 1947) (PDF Edition), 64.

[191] Harding, Christopher. "What Were the Wartime Japanese Thinking?" *History Today*, November 2014, 34.

[192] Beatty, John D., and Lee Rochwerger. *Why The Samurai Lost Japan: A Study in Miscalculation and Folly.* (West Allis, WI, JDB Communications, LLC, 2018)(PDF Edition), 193-195.

www.ingramcontent.com/pod-product-compliance
Lightning Source LLC
Chambersburg PA
CBHW060536130626

46553CB00002B/780